T0329749

FUNDAMENTAL CONCEPTS IN HETEROGENEOUS CATALYSIS

FUNDAMENTAL CONCEPTS IN HETEROGENEOUS CATALYSIS

JENS K. NØRSKOV
FELIX STUDT
FRANK ABILD-PEDERSEN
THOMAS BLIGAARD

WILEY

Published by John Wiley & Sons, Inc., Hoboken, New Jersey
Published simultaneously in Canada

For general information on our other products and services or for technical support, please contact our Customer Care Department within the United States at (800) 762-2974, outside the United States at (317) 572-3993 or fax (317) 572-4002.

Wiley also publishes its books in a variety of electronic formats. Some content that appears in print may not be available in electronic formats. For more information about Wiley products, visit our web site at www.wiley.com.

Library of Congress Cataloging-in-Publication Data:

Nørskov, Jens K.
    Fundamental concepts in heterogeneous catalysis / Jens K. Nørskov, Felix Studt, Frank Abild-Pedersen, Thomas Bligaard.
        pages   cm
    Includes bibliographical references and index.
    ISBN 978-1-118-88895-7 (cloth)
1. Heterogeneous catalysis.   I. Studt, Felix.   II. Abild-Pedersen, Frank.
III. Bligaard, Thomas.   IV. Title.
    QD505.N674 2015
    541′.395–dc23

                                                           2014028859

Printed in the United States of America

10 9 8 7 6 5 4 3 2 1

CONTENTS

Preface   **viii**

1   Heterogeneous Catalysis and a Sustainable Future   **1**

2   The Potential Energy Diagram   **6**

     2.1 Adsorption, 7
     2.2 Surface Reactions, 11
     2.3 Diffusion, 13
     2.4 Adsorbate–Adsorbate Interactions, 15
     2.5 Structure Dependence, 17
     2.6 Quantum and Thermal Corrections to the Ground-State
         Potential Energy, 20

3   Surface Equilibria   **26**

     3.1 Chemical Equilibria in Gases, Solids, and Solutions, 26
     3.2 The Adsorption Entropy, 31
     3.3 Adsorption Equilibria: Adsorption Isotherms, 34
     3.4 Free Energy Diagrams for Surface Chemical Reactions, 40
     Appendix 3.1 The Law of Mass Action and the Equilibrium
                      Constant, 42
     Appendix 3.2 Counting the Number of Adsorbate Configurations, 44
     Appendix 3.3 Configurational Entropy of Adsorbates, 44

4   Rate Constants   **47**

4.1 The Timescale Problem in Simulating Rare Events, 48
4.2 Transition State Theory, 49
4.3 Recrossings and Variational Transition State Theory, 59
4.4 Harmonic Transition State Theory, 61

5   Kinetics   **68**

5.1 Microkinetic Modeling, 68
5.2 Microkinetics of Elementary Surface Processes, 69
5.3 The Microkinetics of Several Coupled Elementary
     Surface Processes, 74
5.4 Ammonia Synthesis, 79

6   Energy Trends in Catalysis   **85**

6.1 Energy Correlations for Physisorbed Systems, 85
6.2 Chemisorption Energy Scaling Relations, 87
6.3 Transition State Energy Scaling Relations in Heterogeneous
     Catalysis, 90
6.4 Universality of Transition State Scaling Relations, 93

7   Activity and Selectivity Maps   **97**

7.1 Dissociation Rate-Determined Model, 97
7.2 Variations in the Activity Maximum with Reaction Conditions, 101
7.3 Sabatier Analysis, 103
7.4 Examples of Activity Maps for Important Catalytic Reactions, 105
     7.4.1 Ammonia Synthesis, 105
     7.4.2 The Methanation Reaction, 107
7.5 Selectivity Maps, 112

8   The Electronic Factor in Heterogeneous Catalysis   **114**

8.1 The d-Band Model of Chemical Bonding at Transition Metal
     Surfaces, 114
8.2 Changing the d-Band Center: Ligand Effects, 125
8.3 Ensemble Effects in Adsorption, 130
8.4 Trends in Activation Energies, 131
8.5 Ligand Effects for Transition Metal Oxides, 134

9   Catalyst Structure: Nature of the Active Site   **138**

9.1 Structure of Real Catalysts, 138
9.2 Intrinsic Structure Dependence, 139
9.3 The Active Site in High Surface Area Catalysts, 143
9.4 Support and Structural Promoter Effects, 146

10   Poisoning and Promotion of Catalysts   **150**

11   Surface Electrocatalysis   **155**

11.1 The Electrified Solid–Electrolyte Interface, 156
11.2 Electron Transfer Processes at Surfaces, 158
11.3 The Hydrogen Electrode, 161
11.4 Adsorption Equilibria at the Electrified Surface–Electrolyte
      Interface, 161
11.5 Activation Energies in Surface Electron Transfer Reactions, 162
11.6 The Potential Dependence of the Rate, 164
11.7 The Overpotential in Electrocatalytic Processes, 167
11.8 Trends in Electrocatalytic Activity: The Limiting Potential Map, 169

12   Relation of Activity to Surface Electronic Structure   **175**

12.1 Electronic Structure of Solids, 175
12.2 The Band Structure of Solids, 179
12.3 The Newns–Anderson Model, 184
12.4 Bond-Energy Trends, 186
12.5 Binding Energies Using the Newns–Anderson Model, 193

Index   **195**

PREFACE

The discovery and development of efficient chemical reactions and processes converting fossil resources into a broad range of fuels and chemicals is one of the most significant scientific developments in chemistry so far. The key to efficient chemical processes is the control of the rates of reaction. This control is usually provided by a catalyst—a substance that can facilitate a chemical reaction and determine the product distribution. The science of catalysis is the science of controlling chemical reactions.

There are many challenges to the science of catalysis that need to be met over the coming years. A sustainable future calls for the development of catalytic processes that do not rely on a net input of fossil resources. This can only be achieved if we discover new catalysts that can efficiently utilize the energy input from the sun or other sustainable sources to synthesize fuels as well as base chemicals for the production of everything from plastics to fertilizers. It also requires more selective processes with fewer waste products and catalysts made from Earth-abundant elements. This represents a formidable challenge. This textbook describes some of the fundamental concepts that will be needed to address this challenge.

Our basic assumption is that the discovery of new catalysts can be accelerated by developing a framework for understanding catalysis as a phenomenon and by pinpointing what are the most important parameters characterizing the chemical properties of the catalyst. We will concentrate in this book on heterogeneous catalysis, that is, catalysts where the processes take place at the surface of the solid. We will develop a systematic picture of the surface-catalyzed processes from the fundamental link to surface geometry and electronic structure to the kinetics of the network of elementary reactions that constitute a real catalytic process. The end result is a theory of variations in catalytic activity and selectivity from one catalyst to the next that will allow the reader to understand the present literature and to make predictions of new catalysts. The latter is aided by the consistent involvement of public databases of surface chemical processes.

The text is aimed at senior undergraduate and graduate students but should be a good guide for any researcher interested in the science and technology of heterogeneous catalysis.

We are grateful to a large number of colleagues for providing material for the book and for providing important feedback during the writing: Rasmus Brogaard, Karen Chan, Søren Dahl, Lars Grabow, Jeff Greeley, Heine Hansen, Anders Hellmann, Jens Strabo Hummelshøj, Karoliina Honkala, Tom Jaramillo, Hannes Jonsson, John Kitchin, Adam Lausche, Nuria Lopez, Nenad Markovic (for providing the unpublished results in Figure 11.10), Andrew Medford, Poul-Georg Moses, Anders Nilsson, Hirohito Ogasawara, Andrew Peterson, Jan Rossmeisl, Jens Sehested, Venkat Viswanathan, Aleksandra Vojvodic, and Johannes Voss. Our thoughts and ideas of this book has been influenced by an additional number of people including Flemming Besenbacher, Charles Campbell, Ib Chorkendorff, Bjerne Clausen, Claus Hviid Christensen, Bjørk Hammer, Karsten Jacobsen, Bengt Kasemo, Norton Lang, Bengt Lundqvist, Alan Luntz, Manos Mavrikakis, Horia Metiu, Yoshitada Morikawa, Matt Neurock, Lars Pettersson, Jens Rostrup-Nielsen, Robert Schlögl, Henrik Topsøe, and Art Williams. We are also grateful to the students of the early versions of the course "Basic Principles of Heterogeneous Catalysis with Applications in Energy Transformations" at Stanford University for many good suggestions.

Stanford, July 2014

Jens K. Nørskov
Felix Studt
Frank Abild-Pedersen
Thomas Bligaard

1

HETEROGENEOUS CATALYSIS AND A SUSTAINABLE FUTURE

The processes that convert fossil resources into fuels and chemicals are essential to modern life. It is, however, also clear that these technologies result in an increased stress on the environment. Even the most efficient processes today result in pollution by by-products. While many chemical production processes have become "cleaner" over the past few decades, the world's consumption of fossil carbon resources has continued to increase. This has resulted in a sharp increase in atmospheric carbon dioxide levels, and because carbon dioxide is a greenhouse gas, the anthropogenic CO_2 emissions have been linked to global climate changes, increased temperatures, melting of the glaciers on all continents, rising sea water levels in the oceans, and the observation of more extreme weather variations across the globe. Since the global population is rapidly growing and many countries are becoming increasingly industrialized, the global energy demand will continue to rise over the next century.

There is a growing consensus that the world's increased demand for fuels and base chemicals will need to be met by more so-called "carbon-neutral" technologies. This calls for new catalytic processes and for catalytic technologies that focus on prevention rather than on remediation.

One central sustainable energy source, which we need to harvest much more efficiently and at a much larger scale than we do today, is sunlight. The annual global energy consumption could be covered by the sunlight striking the Earth within about 1 h assuming that the energy could be efficiently harvested. Consider therefore the

Fundamental Concepts in Heterogeneous Catalysis, First Edition. Jens K. Nørskov,
Felix Studt, Frank Abild-Pedersen and Thomas Bligaard.
© 2014 John Wiley & Sons, Inc. Published 2014 by John Wiley & Sons, Inc.

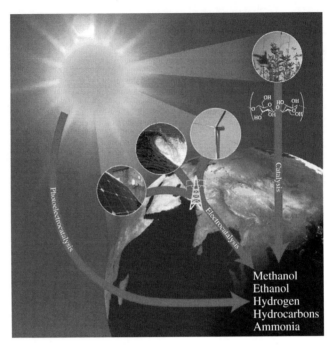

FIGURE 1.1   Illustration of the role of catalysis in providing sustainable routes to fuels and base chemicals. Whether the energy flux from sunlight is harvested through biomass, through intermediate electricity production from photovoltaics or wind turbines, or directly through a photoelectrochemical reaction, the process always requires an efficient catalyst, preferably made of earth-abundant materials. Taken from Nørskov and Bligaard (2013) with permission from Wiley. (*See insert for color representation of the figure.*)

challenge of turning the energy from sunlight into transportation fuels or base chemicals for industry (see Fig. 1.1). Irrespective of whether the sun's energy is harvested by photovoltaic cells, through the use of biomass, wind turbines, wave energy converters, or photoelectrochemical cells, one or more catalysts are needed in order to transform the harvested energy into a useful fuel or chemical. If the goal is to substitute a significant fraction of the global transportation fuel or of base chemicals for industry, the catalysts involved have to be made from elements that are abundant enough that large-scale implementation of the technology can be carried out at a reasonable level of resource utilization and cost.

Traditionally, the field of catalysis is divided into three areas: heterogeneous, homogeneous, and enzyme catalysis. Heterogeneous catalysts are present in a phase different from that of the reactants; typically, the reactants are in the gas or liquid phase, whereas the catalyst is a solid material. Homogeneous catalysts operate in the same phase as the reactants, and enzyme catalysts are specialized proteins. The chemically active part of enzymes is often a tiny part of the protein, and enzyme catalysis can be viewed as a special kind of heterogeneous catalysis.

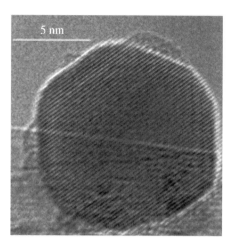

FIGURE 1.2 High-resolution transmission electron microscopy image of a supported Ru catalyst for ammonia synthesis recorded at 552°C and 5.2 mbar in a gas composition of 3:1 H_2/N_2. A Ru particle with a well-formed lattice and surface facets is seen on an amorphous support consisting of BN. A Ba–O promoter phase is observed on top of the Ru particle. Taken from Hansen et al. (2001) with permission from The American Association for the Advancement of Science.

Heterogeneous catalysts have the desirable property that after reaction they are easily separated from the reactants and products. This is an important reason why heterogeneous catalysts are often preferred in industry, in particular for high-volume products, for instance, in the energy sector. For heterogeneous catalysts, the chemical reactions take place at the surface of the material. For that reason, heterogeneous catalysts are typically extremely porous materials so that the surface area is large. In some cases, the catalytic material itself can be made with a high surface area. In other cases, a relatively inert material, the support, is used to stabilize nanoparticles (2–20 nm) of the active material (Fig. 1.2).

Homogeneous catalysts are typically relatively small molecules that are dissolved in the same solution as the reactants and products. Molecular catalysts are often simpler to study, since the active sites on the catalytic molecules can be synthesized with atomic-scale accuracy, and a very detailed understanding of many homogeneously catalyzed processes has therefore been developed.

The focus in the present textbook will be on the fundamental concepts that are needed to understand how solid surfaces act as catalysts. We will introduce a molecular-level understanding of the way surfaces catalyze chemical reactions, which allows the reader to understand why one material is a better catalyst than another for a given reaction. The aim is not to give a complete overview of the types of catalysts or catalytic processes or to give a detailed introduction to the experimental and computational methods that are used to study them. A number of recent textbooks cover these areas very well; see the "Further Reading" list at the end of the chapter. We will use a number of simple catalytic processes as examples throughout but only in order to develop the general rules according to which heterogeneous catalysis works.

SOLAR FUELS

Imagine that we could use renewable electricity, which could come from any of several sources (hydro, solar, wind, geothermal, and others), to directly reduce CO_2 to hydrocarbons and water. Then, we would have a renewable source of fuels for the transportation sector as well as a way of storing energy from intermittent resources. The problem is that there is presently no known catalyst that can do this efficiently. Metallic copper has been demonstrated to produce high (>50%) yields of hydrocarbons at reasonably high ($5\,mA\,cm^2$) current densities (Hori, 2008). But the electrochemical potential needed to make the process run is prohibitively high.

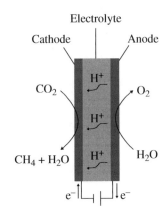

Electrical potential created by photon or from a renewable electricity source.

It turns out that each of the 8 electrons needed to reduce a CO_2 molecule to the simplest hydrocarbon, CH_4 ($CO_2 + 8(H^+ + e^-) \rightarrow CH_4 + 2H_2O$), need on the order of 1 V in extra potential relative to what is needed from a purely thermodynamic point of view in order to make the process run (Kuhl et al., 2012). That means that 8 eV per CO_2 molecule or ~800 kJ/mol is lost in the electrocatalytic reduction of CO_2. A much better catalyst is clearly needed.

REFERENCES

Hansen TW, Wagner JB, Hansen PL, Dahl S, Topsøe H, Jacobsen CJ. Atomic-resolution in situ transmission electron microscopy of a promoter of a heterogeneous catalyst. Science 2001;294:1508.

Hori Y. The catalyst genome. In: Nørskov B, editor. *Modern Aspects of Electrochemistry*, Vol. 42. New York: Springer; 2008. p 89–189.

Kuhl KP, Cave ER, Abram DN, Jaramillo TF. New insights into the electrochemical reduction of carbon dioxide on metallic copper surfaces. Energ Environ Sci 2012;5:7050.

Nørskov JK, Bligaard T. The catalyst genome. Angew Chem Int Ed 2013;52:776.

FURTHER READING

Bruijnincx PCA, Weckhuysen BM. Shale gas revolution: An opportunity for the production of biobased chemicals? Angew Chem 2013;52:11980.

Chorkendorff I, Niemantsverdriet H. *Concepts of Modern Catalysis and Kinetics*. Weinheim: Wiley-VCH Verlag GmbH; 2003.

Ertl G. *Reactions at Solid Surfaces*. Hoboken: John Wiley & Sons, Inc.; 2009.

Nilsson A, Pettersson LGM, Nørskov JK, editors. *Chemical Bonding at Surfaces and Interfaces*. Amsterdam: Elsevier; 2008.

Richter B, editor. *Beyond Smoke and Mirrors: Climate Change and Energy in the 21st Century*. New York: Cambridge University Press; 2010.

van Santen RA, Neurock M. *Molecular Heterogeneous Catalysis*. Weinheim: Wiley-VCH Verlag GmbH; 2006.

Somorjai GA, Li Y. *Introduction to Surface Chemistry and Catalysis*. 2nd ed. Hoboken: John Wiley & Sons, Inc.; 2010.

Thomas JM, Thomas W-J. *Principle and Practice of Heterogeneous Catalysis*. Weinheim: Wiley-VCH Verlag GmbH; 1997.

Yates JT Jr. *Experimental Innovations in Surface Science: A Guide to Practical Laboratory Methods and Instruments*. New York: AIP-Press; 1997.

2

THE POTENTIAL ENERGY DIAGRAM

The central theme in catalysis is the effect of the catalyst on the rate of a chemical reaction or on the product distribution, which is given by the relative rates of different reaction pathways. You can say that catalysis is all about what determines the chemical kinetics. A good catalyst is typically one that gives a high rate and a high selectivity toward the desired product. The reaction rate constant, k, for an elementary reaction is often written as an Arrhenius expression in terms of a prefactor, v, and an activation energy, E_a:

$$k = v e^{-E_a / k_B T} \qquad (2.1)$$

where k_B is the Boltzmann constant and T is the absolute temperature. Variations in the activation energy, when, for example, one catalyst or reactant is substituted with another or when a reaction proceeds through two different reaction mechanisms, are typically large (0.5–2 eV), while the thermal energy, $k_B T$, is small (typically ranges from $k_B T = 0.0257$ eV at $T = 298$ K to $k_B T = 0.1$ eV at $T = 1160$ K). The rate constant is therefore very sensitive to the size of the activation energy. We will return with a more detailed discussion of how the Arrhenius expression comes about in Chapter 4. For now, it suffices to note that any discussion of reaction rates must start with a discussion of the origin of activation energies. For that reason, the starting point of this textbook is an understanding of the potential energy diagram (PED) for surface chemical transformations.

Fundamental Concepts in Heterogeneous Catalysis, First Edition. Jens K. Nørskov,
Felix Studt, Frank Abild-Pedersen and Thomas Bligaard.
© 2014 John Wiley & Sons, Inc. Published 2014 by John Wiley & Sons, Inc.

R VERSUS k_B

If the reaction and activation energies are expressed per amount of substance (in moles), then the Boltzmann constant k_B (approximately $8.61 \cdot 10^{-5}$ eV/K) should be substituted with the gas constant R (approximately 8.31 J/(mol·K)). Since we take an "atomic-scale" viewpoint throughout the book, expressing energies per atom, molecule, or elementary reaction, we shall be utilizing the Boltzmann constant.

VAN DER WAALS INTERACTIONS

The weak bonding due to induced dipole–induced dipole interactions go by several names. Sometimes, the interaction is referred to as London dispersion forces, and in other parts of the literature, they are called van der Waals forces. We shall here not dwell at the more detailed distinction.

2.1 ADSORPTION

We will start by considering the simplest possible potential energy diagrams: those that describe the elementary step of adsorption of a single atom or molecule on a surface. When an atom or a molecule (the adsorbate) approaches a surface, it will start interacting with the electronic states of the solid. At long distances, weak bonding called physisorption dominates. This is due to van der Waals forces, which are purely quantum mechanical in nature and which are relatively long ranged. They occur due to an attraction between mutually induced dipoles of the electron clouds surrounding the atom or molecule and in the surface. Closer to the surface, when the electron clouds of the adsorbate and surface atoms begin overlapping, chemical bonds may form. This stronger form of adsorption is called chemisorption. The strength of the interaction is measured by the change in potential energy of the system as a function of the distance, z, of the adsorbate above the surface:

$$\Delta E(z) = E_{pot}(z) - E_{pot}(\infty) \tag{2.2}$$

In principle, the adsorption energy can be measured, and there are examples where this has been done, typically by inferring an adsorption energy from the measured rate of desorption (temperature-programmed desorption (TPD)). Another way to more directly measure adsorption energies is to measure the temperature increase of a surface as it becomes covered by adsorbates (calorimetry). There are, however, quite few systematic experimental data available. We have therefore chosen throughout this book to illustrate phenomena in terms of energies that are calculated through an approximate solution of the Schrödinger equation based on Kohn–Sham density functional theory (DFT). While DFT is not always in quantitative agreement with experiment, the values for adsorption energies on transition metal surfaces are typically within 0.1–0.2 eV of experiment in the cases where this has been tested. Trends from one catalyst to the next are usually described much better. The use of theoretical interaction energies allows us to always discuss surface reactions and catalysis in terms of the energetics, that is, at the most fundamental level. We will introduce experimental data where possible to illustrate important phenomena and to place the discussion on a firm experimental footing.

The adsorption energy is measured relative to the situation where the adsorbate is far away from the surface, that is, relative to the energy of the clean surface and the free adsorbate. This convention, which will be used throughout the book, means that negative adsorption energy signifies the formation of a chemical bond (the system being stabilized by formation of the bond). Two examples of potential energy diagrams for adsorption are shown in Figure 2.1.

The potential energy diagrams (PEDs) (e.g., Fig. 2.1) contain significant information about an adsorption system. The minimum value of the PED defines the adsorption energy, since it gives the energy gained by adsorption. The location of the minimum defines the equilibrium distance of the adsorbate above the surface. An argon (Ar) atom has a closed outermost electron shell and therefore typically does not form chemical bonds to a metal surface. It only physisorbs with an energy of approximately −0.1 eV on a close-packed Cu surface. A hydrogen (H) atom, on the other hand, forms a strong chemisorption bond to the same surface. The potential energy

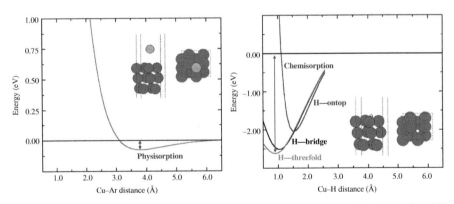

FIGURE 2.1   *Left*: PED for the physisorption of Ar in the threefold position of the Cu(111) surface. The potential energy is shown as a function of the distance between the Cu surface and the adsorbate. The energy of the adsorbate at a distance of 6 Å is chosen as a reference. Due to the filled outermost electronic shell on the Ar atom, this species does not chemisorb to the surface at all, and the shallow physisorption minimum is clearly visible. *Right*: PED for the chemisorption of H on Cu(111) in ontop, bridge, and threefold position.

also depends on the position of the adsorbate parallel to the surface. The most stable position of the H atom is found to be in a site where it has three Cu neighbors. The adsorption energy of −2.6 eV is comparable to the formation energy of an H_2 molecule from H atoms of −2.5 eV per atom. It is thus energetically favorable for an H_2 molecule to dissociate over a Cu(111), which is an important feature in understanding how and why a Cu surface can act as a catalyst, for instance, in the synthesis of methanol from CO and H_2, $2H_2 + CO \rightarrow CH_3OH$.

ENERGY UNITS

Throughout this book, we will be using eV (the kinetic energy gained by an elementary charge accelerated through a potential of 1 V) as the energy unit. It is perhaps not a natural energy unit for chemical processes, since it is neither an SI unit (which would be joule) nor is it the typical atomic unit (hartree). It is, however, a very convenient unit to use when adopting an "atomic-scale" point of view of catalysis. Typical covalent bond strengths in molecules are on the order of 1–10 eV, for instance, and physisorption energies are on the order of 0.1–1 eV. Likewise, a reaction rate of 1 turnover per second corresponds to an activation energy of approximately 0.75 eV at room temperature. In terms of other energy units, 1 eV per atom or molecule = 96.49 kJ/mol, 23.06 kcal/mol, or 0.03675 hartree.

The PED for a molecule approaching a surface is considerably more complicated owing to the fact that it will depend not only on the position of the molecule relative to the surface but also on the intramolecular degrees of freedom as well as the rotational orientation of the molecule relative to the surface. Figure 2.2 shows the two-dimensional potential energy surface (PES), $\Delta E(z,R)$ for H_2, dissociation over Cu(111) as a function of both the distance to the surface, z, and the H–H distance, R, for the molecule positioned parallel to the surface. This PES defines a surface chemical reaction. There are two minima, one where the molecule is far from the surface and one where the molecule has dissociated and the H atoms are well separated and bound to the surface. A considerable energy barrier separates the two minima. This can be observed experimentally by monitoring the probability for dissociation of H_2 molecules scattering off a Cu surface as a function of the kinetic energy of the molecule (see Fig. 2.3). Only molecules with kinetic energies above or just below (due to quantum mechanical tunneling) the energy barrier will be able to dissociate.

It is useful to define the *minimum energy path* (MEP) for a reaction. This minimum path is the lowest energy pathway from one potential minimum on a PES to another (see the box "Minimal Energy Path"). The definition is illustrated in Figure 2.2. The black crosses in the left figure mark the MEP from one local minimum to the other. Plotting the interaction energy as a function of the distance along the path leads to

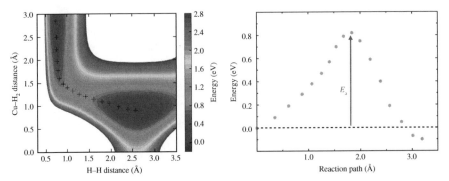

FIGURE 2.2 *Left*: PES for H_2 dissociation over Cu(111). The potential energy of the system is shown as a function of the $Cu–H_2$ and H–H distance, respectively. H_2 far from the Cu surface has been chosen as a reference. The lowest potential energy path for H_2 splitting is marked with black crosses. *Right*: PED for H_2 dissociation where the lowest potential energy (from the figure on the left) is plotted as a function of the reaction path. The PES is calculated without relaxations of the hydrogen and copper atoms. If these are taken into account, a slightly lower barrier of 0.78 eV is found (see CatApp). (*See insert for color representation of the figure.*)

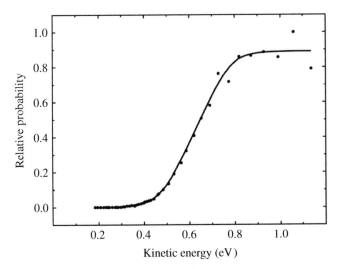

FIGURE 2.3 Measured dissociation probability for a monoenergetic beam of H_2 molecules impinging on a Cu(111) surface as a function of their kinetic energy. Adapted from Rettner et al. (1995).

Figure 2.2 (right). This coordinate is often termed the *reaction coordinate*. The one-dimensional (1D) representation of the potential energy curve along the reaction path is a very convenient way of showing the energy variation of a complete elementary reaction. It directly identifies the activation energy, E_a, in an Arrhenius expression for the reaction rate (see the box "Minimal Energy Path" and Chapter 4).

MINIMAL ENERGY PATH (MEP)

The MEP between two stationary points on the PES can more formally be defined as the continuous and smooth path among all possible paths connecting the two stationary points, which satisfies the two following properties:

1. It is a path of least action: At any point along the path, the gradient of the potential has no component perpendicular to the path.
2. The highest potential energy along the MEP is equal to or lower than the highest potential energy along all stationary paths.

The MEP defined in this way will typically have an associated highest potential energy in a point where the derivative of the potential is zero, both perpendicular to the path and along the path (since the point is a maximum along the path). First, we assume that the second derivative is positive along all directions perpendicular to the path in the highest energy point (since we could otherwise minimize along these directions). Next, we also assume that the second derivative along the path is negative in this point—that the highest energy is not attained in a plateau. Then, we can conclude that the highest energy point along the path is a first-order saddle point. We shall for now take the energy difference between the initial state of the reactant and the first-order saddle point as a measure for the activation energy. In Chapter 4, we shall address through the use of "transition state theory" why this is a reasonable approximation.

2.2 SURFACE REACTIONS

The dissociation of H_2 over a Cu surface discussed earlier is an example of an elementary surface reaction. Other types of elementary surface reactions are illustrated in Figure 2.4. After reactants are adsorbed and perhaps dissociated, they will diffuse and recombine to form new molecules before the product is desorbed into the surrounding gas or liquid phase. For each of these different elementary reaction steps, we can define a PES and a 1D PED.

A large number of such PEDs have been calculated and a web-based database has been created, which can be accessed through an Internet browser. The website http://suncat.slac.stanford.edu/catapp/ features a list of hyperlinks to the available versions of the tool together with a list of references to the scientific data it employs. Using the application, one can choose a surface and an elementary reaction step and be presented with a reaction path reporting the reaction and activation energy. Examples of such 1D PEDs from the *CatApp* are shown in Figure 2.5.

One can combine a series of elementary reactions steps into a PED for a complete catalytic process. Let us take the ammonia synthesis reaction as an example. The overall reaction can be written as

$$N_2 + 3H_2 \rightarrow 2NH_3,$$

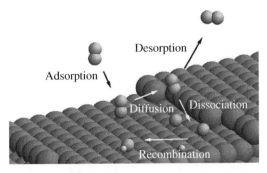

FIGURE 2.4 Illustration of the elementary reaction steps on surfaces. (*See insert for color representation of the figure.*)

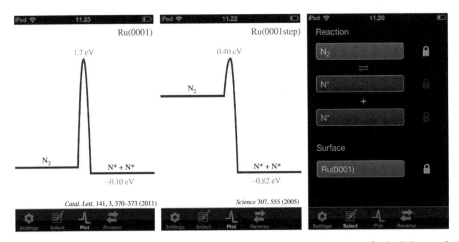

FIGURE 2.5 These screenshots from the CatApp (http://suncat.slac.stanford.edu/catapp/) show examples of elementary reaction PEDs that can be obtained from this tool. *Left* and *center*: N_2 splitting on close-packed and stepped Ru(0001), respectively. *Right*: select view of the CatApp. Here, the user can choose the reaction and surface parameters from drop-down menus. Taken from Hummelshøj et al. (2012) with permission from Wiley.

consisting of the following elementary reaction steps:

1. $N_2 + 2* \rightarrow 2N*$
2. $H_2 + 2* \rightarrow 2H*$
3. $N* + H* \rightarrow NH* + *$
4. $NH* + H* \rightarrow NH_2* + *$
5. $NH_2* + H* \rightarrow NH_3* + *$
6. $NH_3* \rightarrow NH_3 + *$

In Figure 2.6, we show the combined PED for the full ammonia synthesis reaction on a stepped Ru surface. That the initial and final states of the catalyst are the same and the net result is that the overall reaction has run once. The potential energy difference

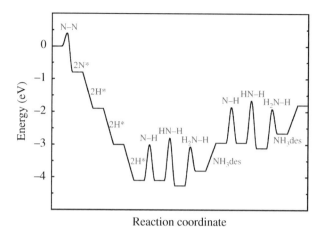

FIGURE 2.6 PED for ammonia synthesis on the stepped Ru(0001). The numbers correspond to the six different reaction steps that are defined earlier. The data for the six reaction steps has been obtained from CatApp and "glued" together to yield the reaction diagram.

between the initial and final states is the reaction energy of ammonia synthesis. In the case of ammonia synthesis, the reaction energy is −1.8 eV. If we take zero-point energy (ZPE) contributions (we will discuss this and other energy contributions in Chapter 3) into account, the reaction energy is −1.00 eV.

The PED for ammonia synthesis illustrates the basic principle by which a catalyst works. N_2 is an extremely stable molecule. It is inert in the gas phase, and for hydrogen to attack it, the N–N triple bond must first be activated or split. To split the triple bond costs 9.76 eV in the absence of a catalyst, and that is why gas-phase ammonia synthesis is only possible at extremely high temperatures such as in lightning or in electrical arcs. The role of the catalyst surface is to provide stabilization of the N atoms during and after dissociation of the molecule. In the presence of a Ru catalyst, there are no activation energies larger than approximately 1 eV. In the presence of a catalyst, ammonia can thus be formed at much more moderate temperatures. We will return to a more quantitative analysis of this example later in the book.

2.3   DIFFUSION

Diffusion is another of the elementary steps in surface chemical reactions, and we can define a PES for such processes as well. Figure 2.7 shows an example for the diffusion of a H atom over the Cu(111) surface. We already established in Figure 2.1 that hydrogen adsorption is strongest in the threefold site, followed by bridge (twofold) and ontop (onefold) adsorption. This can be seen from the PES in Figure 2.7 as well. Importantly, these findings show that hydrogen diffusion from one threefold site to another can easily occur over the bridge site with a very small diffusion barrier on the order of 0.2 eV.

It is generally so that diffusion barriers for simple adsorbates on metal surfaces are not much higher than 0.5 eV. This results in diffusion being very fast at temperatures above 300 K where most industrial catalytic processes take place. If, however, there are high coverages of adsorbates, such that there are no free sites available to diffuse into, this picture may change. It has been found experimentally and by DFT calculations that for adsorption on metal surfaces, activation barriers for diffusion are typically of the order of 10–15% of their adsorption energy. This is illustrated for a range of adsorbates in Figure 2.8. It can be understood as a consequence of the fact that during diffusion the surface–adsorbate bond is only "partially broken," and hence, only a small fraction of the maximum adsorption energy is lost when moving along the diffusion path.

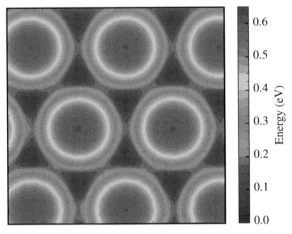

FIGURE 2.7   PES of H diffusion on Cu(111). The potential energy is plotted over part of the Cu surface area. H adsorbed in the threefold position has been chosen as the reference.

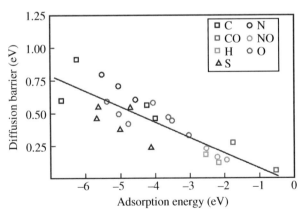

FIGURE 2.8   Diffusion versus adsorption on metal surfaces. The diffusion barriers for a range of different adsorbates are plotted as a function of their adsorption energy. Adapted from Nilekar et al. (2006). (*See insert for color representation of the figure.*)

On metal oxide surfaces and other inorganic compound catalysts, where the distance between adsorption sites are larger than on metal surfaces, diffusion barriers are often found to be larger. This means that the mobility of adsorbates on the surfaces of compounds can become important for the overall reaction rate.

2.4   ADSORBATE–ADSORBATE INTERACTIONS

The adsorption energy will in general depend on the presence of other adsorbates on the surface. Figure 2.9 shows the interaction energy between two O atoms on a Pt surface as a function of the distance between them. The O atoms are always in the same local configuration surrounded by three Pt atoms. When the O atoms get close to each other, so that they bond to the same platinum atoms, their adsorption energies become considerably weaker.

The dominant adsorbate–adsorbate interaction for O atoms on the close-packed Pt surface is the repulsion at short distances. This is typically the case for strongly chemisorbed adsorbates on metal surfaces. A repulsion at short distances means that the adsorption bond becomes weaker with increasing coverage. This is an important contribution to the adsorption energy, which often keeps the coverage of adsorbates limited to less than a monolayer.

If there are N_0 surface sites on a clean surface and we adsorb N adsorbates on that surface, we define the fractional *coverage* as $\theta = N/N_0$. In terms of this coverage, the average adsorption energy is defined as

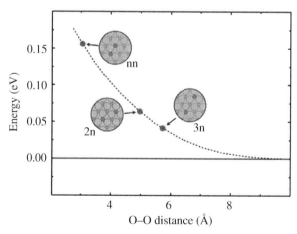

FIGURE 2.9   Interaction energy between two O atoms on a Pt(111) surface as a function of the distance between them. The energy of the oxygen atoms sitting far from each other on Pt(111) has been selected as the energy reference. The surface sites are shown in the inset for nearest neighbor (nn), next nearest neighbor (2n), and third nearest neighbor (3n).

$$\Delta E_{av}\left(\theta\right)=\frac{E_{surf.+N^{*}ads.}-E_{surf.}-NE_{ads.}}{N} \tag{2.3}$$

where the energies entering this expression are for a surface with N adsorbates, $E_{surf.+N^{*}ads.}$; for the clean surface, $E_{surf.}$; and for the adsorbate in vacuum, $E_{ads.}$. Since the average adsorption energy varies with the coverage, we could consider it a result of integrating up a differential adsorption energy, $\Delta E_{diff}(\theta)$, from zero coverage up to the coverage in question: $\Delta E_{av}(\theta)=\theta^{-1}\int_{\theta'=0}^{\theta}\Delta E_{diff}(\theta')d\theta'$. The differential adsorption energy is therefore

$$\Delta E_{diff}\left(\theta\right)=\frac{d\left(N\cdot\Delta E_{av}\right)}{dN}=\frac{d\left(\theta\cdot\Delta E_{av}\right)}{d\theta}. \tag{2.4}$$

ΔE_{diff} measures the adsorption energy of the "last adsorbate." Figure 2.10 shows the variation with coverage of the average and differential adsorption energy of O on a close-packed Pt surface. The adsorption energy increases (the bond becomes weaker) steadily for coverages of 0.25 monolayers and above.

There can also be attractive interactions between adsorbates that are not sitting directly in the vicinity of each other. At low temperatures, this can lead to the formation of ordered patterns. A repulsive interaction can also lead to ordered structures, since the adsorbates form the structure where their mutual distance is maximized at a given coverage. This is illustrated in Figure 2.11 for O adsorption on Pd(111) at 0.25 ML coverage.

Repulsive interactions between two different kinds of adsorbates can also lead to pattern formation. Consider, for instance, CO and O adsorbed on a Pd surface. These could, for example, be found on the surface of a car catalyst where CO is oxidized to CO_2. If the CO–O repulsion is stronger than the O–O and the CO–CO repulsion, then the O atoms and the CO molecules will, at high coverage, tend to clump together in

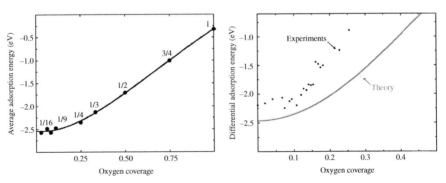

FIGURE 2.10 *Left*: Average adsorption energy of oxygen on Pt(111) as a function of coverage (from theory). *Right*: Measured differential adsorption energy vs. coverage on the Pt(111) surface (black dots). Theoretical differential heats of adsorption as derived from the calculations shown in the left figure is included for comparison (grey line). Figure adapted from Fiorin et al. (2009) and Karp et al. (2012).

FIGURE 2.11 STM images of a Pd(111) surface with an ordered oxygen p(2 × 2) structure (structure with a unit cell that is twice as large as the metal surface unit cell in two directions). Oxygen atoms are imaged as bright bumps. Taken from Méndez et al. (2005) with permission from The American Physical Society.

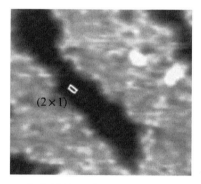

FIGURE 2.12 STM images of a Pd(111) surface with oxygen p(2 × 1) islands and CO-covered surfaces. Oxygen is imaged dark. Pattern formation of oxygen and CO islands can clearly be seen. Taken from Méndez et al. (2005) with permission from The American Physical Society.

separate islands to minimize the length of the boundary where adsorbed CO and O are in touch. This is shown in Figure 2.12.

2.5 STRUCTURE DEPENDENCE

The final theme in this chapter is the surface structure dependence of the adsorption energy and reaction energies and activation barriers. It was illustrated in Figure 1.1 that catalysts are often nanoparticles exposing several different facets to the gas phase. They also have other types of sites such as edges and corners. Figure 2.13 shows a schematic of such a catalyst particle. Facets are usually denoted by their Miller indices. The most common facets are the most close packed, which for the fcc crystal structure is the (111) and the slightly more open (100) surface structure.

In a reactive environment, the reactants and products can induce other facets or completely new structures to become exposed. Undercoordinated sites at edges and corners are often particularly important for catalysis. The same kind of sites is found at steps on the surface. For the fcc structure, a (211) step is often used to model undercoordinated sites.

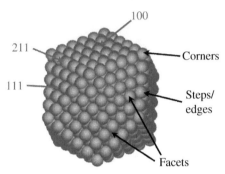

FIGURE 2.13 Schematic picture of a Pt particle consisting of 561 atoms. The particle is shaped so that it consists of (111) and (100) facets that are connected along edges that locally resemble the step sites on a (211) surface.

MILLER INDICES

Miller indices form a notation system in crystallography for directions and planes in crystal lattices. Lattice planes are determined by the three integers h, k, and l, also called Miller indices. In a cubic lattice, these indices coincide with the inverse intercepts along the lattice vectors as shown in Figure 2.14. Thus, (hkl) simply denotes a plane that intercepts at the three lattice vectors at the points a/h, b/k, and c/l (or a multiple of those). If one of the indices is zero, the planes are parallel to that axis.

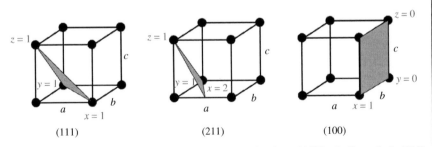

FIGURE 2.14 Three examples showing the determination of Miller indices. *Left:* (111) plane intercepting at 1 for x, y, and z. *Center:* (211) plane intercepting at ½ for x and y and 1 for z. *Right:* (100) plane only intercepting at $x = 1$.

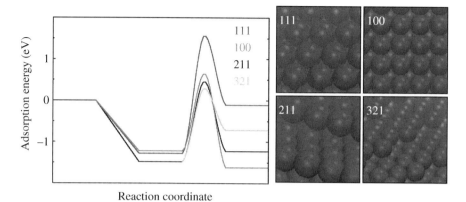

FIGURE 2.15 *Left*: CO dissociation on different facets of fcc Ni. CO is first adsorbed molecularly and then dissociated into adsorbed carbon and oxygen. The barrier for dissociation is extremely structure sensitive being almost an eV higher over the (111) facet than over the more undercoordinated facets. *Right*: illustration of the fcc (111), (100), (211), and (321) facet structures. Adapted from Andersson et al. (2008). (*See insert for color representation of the figure.*)

The PED for elementary surface reactions generally depends strongly on the surface structure. We illustrate this in Figure 2.15 by comparing the PED for CO dissociation on different Ni facets. Clearly, steps on the surface interact stronger with the final product, adsorbed C and O. The dissociation barrier is also considerably lower. This means that not all sites on a metal nanoparticle will contribute equally to the catalytic activity. Any process involving CO dissociation will, for instance, be strongly favored at defects and at the edges and corners of the particles.

When metallic nanoparticles become very small, there are additional effects due to the finite size of the particle. In the limit of particles consisting of just a few metal atoms, the catalyst is not a metal any more—rather, it is a molecule with properties that are quite different from metallic surfaces. The transition from metal to molecule appears to happen at rather small particles that are about 2 nm in diameter. This is illustrated in Figure 2.16 where the oxygen adsorption energies on the (111) and (211) facets of Pt nanoparticles are plotted as a function of their particle size. It can be seen that the oxygen binding energies approach those found for the (111) and (211) facets of infinite sizes for particles that are larger than 2 nm in diameter.

For oxides and other nonmetallic catalysts, there could be considerably stronger effect of size. This is a problem that remains to be solved. We also note that there are additional structural effects due to the interaction between the active phase and the support—in some cases, they work together to form a new active phase at the boundary. We will return to this question in Chapter 9.

SIGNIFICANCE OF STRUCTURAL ACTIVATION ENERGY DIFFERENCES FOR DISSOCIATION REACTIONS

The activation energy differences of about 1 eV observed in Figure 2.15 are important for two reasons. One reason is that a 1 eV barrier difference leads to large differences in the rates of the elementary reactions over the different facets. The other reason is that this observation is general. We shall in later chapters discuss further how such huge activation energy differences are present for many dissociation reactions of strongly bonded molecules and fragments over transition metals.

To make a rough quantitative estimate of the difference in catalytic activity between the close-packed and the undercoordinated sites of Ni for processes involving CO dissociation, we shall make a couple of assumptions. Let us assume that CO dissociation is the important rate-limiting step of the process, such that the catalytic rate is entirely determined by the rate constant for CO dissociation. This assumption may or may not be correct in general, but if the barriers become high enough, it will be a reasonable assumption. Now, let us further assume that the prefactors for the rates are approximately equal over the different surfaces. This is typically a quite reasonable first approximation. Finally, let us assume that coverages do not play a major role in determining the catalytic rates and that the process we are thinking of runs at 580 K where $k_B T = 0.05$ eV. Then, we would expect the undercoordinated surfaces to be half a billion times faster at this catalytic reaction than the close-packed (111) surface. We shall in later chapters go to some length toward developing a more accurate quantitative treatment of the catalytic rates by relaxing the assumptions made earlier.

2.6   QUANTUM AND THERMAL CORRECTIONS TO THE GROUND-STATE POTENTIAL ENERGY

We end this chapter by discussing a couple of relevant quantum and thermal corrections to the potential energy.

A classical mechanical system at equilibrium is at rest in a (local) minimum on the PES and will have a potential energy given by the minimum value, E_{pot}^{min}. A (real) quantum system will not, in general, be able to attain such an absolute potential energy minimum. This is due to the Heisenberg uncertainty principle, which prescribes that the uncertainty in the position of a particle, Δx, is related to the uncertainty in the momentum, Δp, of the particle:

$$\Delta x \cdot \Delta p \geq \frac{\hbar}{2} \tag{2.5}$$

This prevents the particle from being perfectly at rest in a perfectly specified position. For a bound system, such as an adsorbate, small fluctuations around the local potential energy minimum are often accurately represented as harmonic vibrations. This allows us to make a simple correction to the potential energy minimum in order to obtain a

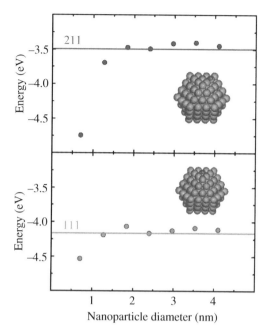

FIGURE 2.16 Adsorption energies of oxygen on platinum nanoparticles as a function of their diameter. Oxygen adsorption is shown for adsorption on the (211) edge (dots, top) and the (111) facet (dots, bottom). Oxygen adsorptions on the (211) (line, top) and (111) (line, bottom) facet of a particle with infinite diameter (metallic properties) are shown for comparison. Adsorption of oxygen on a platinum particle with a diameter of ~2 nm is shown schematically for the (111) facet and (211) bridge position. It can be seen that adsorption energies converge to metal values for particles that have diameters around 2 nm. Figure adapted from Li et al. (2013).

reasonable estimate of the ground-state energy of a quantum system. The quantized energy solutions, E_n, of the harmonic oscillator quantum problem are given by

$$E_n = E_{pot}^{min} + \left(n + \frac{1}{2}\right)h\nu_i \tag{2.6}$$

where E_{pot}^{min} is the local minimum in the potential, ν_i is the vibrational frequency, h is Planck's constant, and n is a quantum number. The ground state, which is the lowest energy level, $n=0$, thus has an energy given by

$$E_0 = E_{pot}^{min} + \frac{1}{2}h\nu_i \tag{2.7}$$

The contributions from different vibrational modes are additive, such that if there are N_{modes} vibrational modes in the potential energy minimum, the total so-called *ZPE* correction is given by

$$ZPE = \sum_i^{N_{modes}} ZPE^i = \sum_i^{N_{modes}} \frac{1}{2}h\nu_i \tag{2.8}$$

Figure 2.17 shows how the *ZPE* for one vibrational mode depends linearly on the vibrational frequency of, for example, a molecule. Importantly, frequencies that are significantly smaller than $1000\,cm^{-1}$ contribute only little to *ZPE*, while frequencies larger than $1000\,cm^{-1}$ can contribute several tenths of eVs. We shall see later that this trend is completely opposite to that of the frequency contribution to the entropy (see Fig. 3.2). While most frequencies of adsorbed molecules are below $1000\,cm^{-1}$, frequencies that involve bond stretching of strongly bound atoms usually end up in the region above $1000\,cm^{-1}$ (the internal N_2 stretching frequency is, for instance, $2358\,cm^{-1}$). The H_2 frequency is as high as $4395\,cm^{-1}$, and its contribution is therefore quite significant. As we will show here, the inclusion of *ZPE* contributions is especially important for hydrogenation reactions, such as ammonia synthesis. The reason lies in the multiple new atom–H stretching frequencies that are created in the hydrogenation of molecules and atoms (e.g., the three N–H stretching frequencies in NH_3). The new frequencies are often well above $1000\,cm^{-1}$ and thus contribute significantly to the *ZPE* and usually more than make up for the loss in *ZPE* that occurs due to the breaking of the H–H (and N–N) bond. Table 2.1 shows the reaction *ZPE* corrections for three simple hydrogenation reactions. The reaction energy for ammonia synthesis, for example, comes out as much as 0.83 eV too exothermic if one does not account for the *ZPE* corrections.

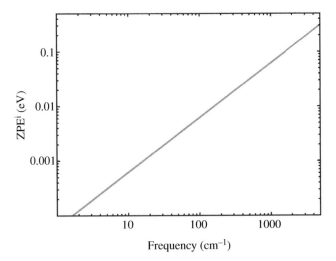

FIGURE 2.17   *ZPE^i* plotted as a function of vibrational frequency. Note that the correlation is linear and plotted on a double logarithmic scale.

TABLE 2.1   ZPE correction for selected hydrogenation reactions

Reaction	ZPE correction (eV)
$N_2 + 3H_2 \rightarrow 2NH_3$	0.83
$CO + 3H_2 \rightarrow CH_4 + H_2O$	0.80
$CO + 2H_2 \rightarrow CH_3OH$	0.71

Temperature is a measure of the average kinetic energy of the atoms in a system at thermal equilibrium. As the temperature increases, the kinetic energy will increase proportionally. This means that at nonzero temperatures the system will contain more energy than the (ZPE-corrected) potential energy, which is calculated at $T=0\,\mathrm{K}$. Classically, we say that in order to "heat up" a system, we need to transfer energy to the system, such that the atoms move around and vibrate faster, thus increasing the kinetic and (often also) the potential energy of the atoms making up the system. The amount of energy that needs to be transferred per temperature increase at a given temperature in order to warm up the system is the heat capacity (at constant pressure), $C_p(T)$. The sum of the (ZPE-corrected) ground-state energy, E_0, and the thermal energy of the system shall be referred to as the system's internal energy, $U(T)$. If we know how the heat capacity $C_p(T)$ varies with temperature, we can find the internal energy by integrating $C_p(T)$ up to the relevant temperature:

$$U(T) = E_0 + \int_{T=0}^{T} C_p(T')dT' \tag{2.9}$$

Typically $C_p(T)$ is small enough such that the second term on the right side of Equation (2.9) is a small correction to the first.

QUANTUM EXPRESSION FOR THE INTERNAL ENERGY

In a quantum formulation, the variation of the internal energy comes as a variation in which excited quantum states are thermally occupied at different temperatures. At $T = 0$ K, only the ground state will be occupied. As the temperature increases, some of the other low-lying quantum states become occupied. Fundamental in the quantitative analysis of thermal properties is the *Boltzmann distribution*, which specifies that the relative probability of finding the system (which is in thermal contact with a heat reservoir) in two different quantum states is described entirely by the energy difference between the states and the system temperature through the relation:

$$\frac{P(E_i)}{P(E_j)} = e^{-\frac{(E_i - E_j)}{k_B T}} \tag{2.10}$$

At nonzero temperature, the potential energy is therefore not E_0, but a suitable average over many different quantum states, with a higher weight on low-energy states. The useful tool in analyzing a system in thermal contact with a heat reservoir (a canonical system) is the *canonical partition function*:

$$Z(T) = \sum_{i=0}^{\infty} e^{-\frac{E_i}{k_B T}} \tag{2.11}$$

The partition function gives a simple relation for determining the absolute probability for finding the system in a given quantum state, i:

$$P_i = \frac{e^{-\frac{E_i}{k_B T}}}{Z} \tag{2.12}$$

This is now a correctly normalized absolute probability because the relative probabilities are conserved and the sum of probabilities for finding the system in any state is unity:

$$\sum_{i=0}^{\infty} P_i = \frac{\sum_{i=0}^{\infty} e^{-\frac{E_i}{k_B T}}}{Z} = \frac{Z}{Z} = 1 \tag{2.13}$$

This means that we can arrive at a quantum statistical mechanics definition of the internal energy as the thermally averaged energy of the various occupied states:

$$U(T) = \langle E \rangle = \sum_{i=0}^{\infty} P_i E_i = \frac{\sum_{i=0}^{\infty} E_i e^{-\frac{E_i}{k_B T}}}{Z} \tag{2.14}$$

REFERENCES

Andersson MP, Abild-Pedersen F, Remediakis IN, Bligaard T, Jones G, Engbæk J, Lytken O, Horch S, Nielsen JH, Sehested J, Rostrup-Nielsen JR, Nørskov JK, Chorkendorff I. Structure sensitivity of the methanation reaction: H_2-induced CO dissociation on nickel surfaces. J Catal 2008;255:6.

Fiorin V, Borthwick D, King DA. Microcalorimetry of O_2 and NO on flat and stepped platinum surfaces. Surf Sci 2009;603:1360.

Hummelshøj JS, Abild-Pedersen F, Studt F, Bligaard T, Nørskov JS. CatApp: a web application for surface chemistry and heterogeneous catalysis. Angew Chem Int Ed 2012;51:272.

Karp EM, Campbell CT, Studt F, Abild-Pedersen F, Nørskov JK. Energetics of oxygen adatoms, hydroxyl species and water dissociation on Pt(111). J Phys Chem C 2012; 116:25772.

Li L, Larsen AH, Romero NA, Morozov VA, Glinsvad C, Abild-Pedersen F, Greeley J, Jacobsen KW, Nørskov JK. Investigation of catalytic finite-size-effects of platinum metal clusters. J Phys Chem Lett 2013;4:222.

Méndez J, Kim SH, Cerdá J, Wintterlin J, Ertl G. Coadsorption phases of CO and oxygen on Pd(111) studied by scanning tunneling microscopy. Phys Rev B 2005;71:085409.

Nilekar AU, Greeley J, Mavrikakis M. A simple rule of thumb for diffusion on transition-metal surfaces. Angew Chem Int Ed 2006;45:7046.

Rettner CT, Michelsen HA, Auerbach DJ. Quantum-state-specific dynamics of the dissociative adsorption and associative desorption of H_2 at a Cu(111) surface. J Chem Phys 1995; 102:4625.

FURTHER READING

Ashcroft NW, Mermin DN. Crystal lattices (Chapter 4) & The reciprocal lattice (Chapter 5). In: *Solid State Physics*. Philadelphia: Saunders College Publishing; 1976.

Burke and friends. 2007. The ABC of DFT. Available at http://dft.uci.edu/doc/g1.pdf. Accessed May 29, 2014.

Hummelshøj JS, Abild-Pedersen F, Studt F, Bligaard T, Nørskov JS. CatApp: A web application for surface chemistry and heterogeneous catalysis. Angew Chem Int Ed 2012;124:278. Available at http://suncat.slac.stanford.edu/catapp/. Accessed May 29, 2014.

Jónsson H, Mills G, Jacobsen KW. Nudged elastic band method for finding minimum energy paths of transitions. In: Berne BJ, Ciccotti G, Coker DF, editors. *Classical and Quantum Dynamics in Condensed Phase Simulations*. Singapore: World Scientific; 1998.

Kohn W. Electronic structure of matter-wave functions and density functionals. Rev Mod Phys 1999;71:1253.

Miller SD, Kitchin JR. Relating the coverage dependence of oxygen adsorption on Au and Pt fcc(111) surfaces through adsorbate-induced surface electronic structure effects. Surf Sci 2009;603:794.

Silbey RJ, Alberty RA, Bawendi MG. Statistical mechanics. In: *Physical Chemistry*. Hoboken: John Wiley & Sons, Inc; 2005.

Wu C, Schmidt DJ, Wolverton C, Schneider WF. Accurate coverage—Dependence incorporated into first—Principles kinetic models: Catalytic NO oxidation on Pt(111). J Catal 2012;286:88.

3

SURFACE EQUILIBRIA

As an important step toward understanding rates of chemical reactions on catalyst surfaces, we need first to understand equilibria involving atoms and molecules bound to a surface. In fact, an understanding of the stability of adsorbed intermediates in a surface-catalyzed reaction can in many cases give a good indication of whether the catalyst will work or not, even before analyzing the reaction barriers involved in the problem. In this chapter, we start by giving a brief review of the theory of chemical equilibria. We then extend those concepts to describe surface reactions.

3.1 CHEMICAL EQUILIBRIA IN GASES, SOLIDS, AND SOLUTIONS

The key thermodynamic concept for describing equilibria in chemical processes is the Gibbs free energy. The Gibbs free energy is defined as

$$G = H - TS, \qquad (3.1)$$

where S is the entropy and H is the enthalpy, which is defined from the internal energy, U, as

$$H = U - pV, \qquad (3.2)$$

Fundamental Concepts in Heterogeneous Catalysis, First Edition. Jens K. Nørskov,
Felix Studt, Frank Abild-Pedersen and Thomas Bligaard.
© 2014 John Wiley & Sons, Inc. Published 2014 by John Wiley & Sons, Inc.

where p is the pressure and V is the volume. For a very short discussion on entropy, please see Appendices 3.2 and 3.3. The Gibbs free energy plays a role in chemistry much like the role played by the potential energy in classical mechanics. The potential energy describes a mechanical system's potential for carrying out mechanical work, and the Gibbs free energy describes a closed chemical system's potential for carrying out nonexpansion work. The maximum amount of work carried out by a mechanical system is only attained if all its motion is frictionless; likewise, the maximum amount of nonexpansive work carried out by a chemical system can only be attained if all reactions are reversible.

A mechanical system is stable when it is stationary in its lowest attainable potential energy configuration; a chemical system is in equilibrium when it is in its lowest attainable Gibbs free energy configuration. For a mechanical system, this equilibrium occurs where the derivative of the potential energy (or the negative force) is zero with respect to variations along all the system's degrees of freedom. For a chemical system, it occurs when the variation of the Gibbs free energy (at constant temperature and total pressure) with respect to all variations in composition of the system is zero.

The practical utilization of the concept of Gibbs free energy is therefore that equilibrium of the system requires that any possible (element-conserving) chemical reaction that the system could undertake must satisfy the relation

$$\Delta G = G_{\text{final}} - G_{\text{initial}} = 0 \tag{3.3}$$

where G_{initial} and G_{final} are the Gibbs free energies of a given set of atoms before and after, respectively, they undergo some chemical reaction.

Generally, S and U depend on the temperature and so does H. The Gibbs free energy, in contrast to what Equations (3.9) and (3.10) would seem to suggest at a first glance (see following text), ends up depending nonlinearly on the temperature. The entropy can be thought of as a measure of the number of accessible quantum states of a system (see Appendices 3.2 and 3.3), and the temperature dependence comes from extra quantum states becoming accessible as the temperature increases. (Classically, one would say that the increased kinetic energy of the atoms allows them to sample higher potential energy regions of the potential energy surface, whereas quantum mechanically one would say that higher energy levels become increasingly occupied). We saw that the temperature dependence of the internal energy can be taken into account by including an integral of the heat capacity over temperature, and a similar expression can be written for the entropy. We shall for the most part (for nonadsorbates) neglect the detailed temperature dependence, since it is typically easily retrieved from standard tables.

The entropy per atom or molecule of an ideal gas or a dilute solution is given by expressions such as

$$S_{\text{gas}}(p) = -k_{\text{B}} \ln p \tag{3.4}$$

$$S_{\text{solute}}(C) = -k_{\text{B}} \ln C \tag{3.5}$$

The pressure (p) and concentration (C) dependence basically stems from the fact that the entropy is proportional to the logarithm of the number of accessible states, Ω, as

$$S = k_B \cdot \ln \Omega \qquad (3.6)$$

Since the number of accessible quantum states of a randomly moving noninteracting entity (such as a molecule in a dilute gas or an ion in dilute solution) is proportional to the volume per particle, which basically is given by the number of thermally accessible quantum states for a free particle in a box, the number of quantum states becomes inversely proportional to the pressure or concentration.

Reactions involving liquid or solid phases result in changes in volume at fixed density, as opposed to the changes in particle densities at fixed volume in the case of gases or ions in solution. This means that the Gibbs free energies of liquids and solids do not have terms varying with the logarithm of their particle densities (i.e., the free energies of solids and liquids do not have terms analogous to Equations 3.4 and 3.5).

The absolute Gibbs free energy incorporates a certain level of arbitrariness through the dependence of the internal energy on the ground-state potential energy. The commonly accepted standardized choice is to define a *standard state* for every substance at a set of *standard conditions* for every type of substance and a reference state for every element. The standard condition for a gas is chosen to be $p° = 1$ bar, and the standard condition for a solution is the standard concentration molarity of $C° = 1$ mol solute/kg solvent at the standard pressure of $p° = 1$ bar. The standard conditions for liquids and solids are also the standard pressure of $p° = 1$ bar. (But since their free energy varies very little with pressure, this choice of standard is less important for the applications we shall discuss.) The standard state of a substance is chosen as its most stable thermodynamic state under standard conditions. The reference state for every element is then defined as its most stable pure standard state at standard conditions. With these choices, one can define the standard enthalpy of formation, $H°$, for each element as being zero for the element in its reference state, while for other substances $H°$ is the reaction enthalpy for making the substance from the constituting elements in their reference state. Similarly, the standard entropy, $S°$, is then defined as the entropy of a given substance at standard conditions.

For historical reasons, $G°$, is also defined to be zero for the pure elements in their reference states. For other substances, $G°$ is the reaction Gibbs energy for making the substances in their standard state from the constituting elements in their reference states. This is a point that one should be aware of, because its implication is that when *using tabulated values*, generally

$$G° \neq H° - TS° \quad \text{and} \quad G \neq H - TS \qquad (3.7)$$

The reason that the nonessential zero definition of the Gibbs energy has been made, even if it "messes up" one of the central expressions in physical chemistry, is that one typically always needs the Gibbs energies of reaction, not the absolute Gibbs energies themselves. The following relation will therefore still hold for the Gibbs energies of reaction:

$$\Delta G° = \Delta H° - T\Delta S° \tag{3.8}$$

The Δ's here indicate a change in the given quantity as a reaction is undertaken. The definition of zero standard Gibbs energies for pure substances in their reference state makes it somewhat faster to look up reaction free energies in tables of thermodynamic data, since one will not necessarily need to look up $G°$ for pure substances.

For electrochemical reactions, the reactivity depends on variations in the electrostatic potential. The absolute potential is often defined as zero for an electrode over which H_2 (gas) is in equilibrium with a 1 M acid solution. (Such a solution has a pH of 0, since pH is the negative of the logarithm of the concentration deviation of H^+ ions from the standard concentration.) This definition of an electrochemical potential thus couples the standards of concentrations of solutions and pressures of gases with the definition of a potential in which equilibrium is established.

For a pure substance, X, we can write the following:

$$G^X_{(g)}(p) = G^X_{(g)}° + (p - p°)v + k_B T \cdot \ln\left(\frac{p}{p°}\right) \tag{3.9}$$

$$G^X_{(aq)}(C, p) = G^X_{(aq)}° + (p - p°)v + k_B T \cdot \ln\left(\frac{C}{C°}\right) \tag{3.10}$$

$$G^X_{(l)}(p) = G^X_{(l)}° + (p - p°)v \tag{3.11}$$

$$G^X_{(s)}(p) = G^X_{(s)}° + (p - p°)v \tag{3.12}$$

where the $(p - p°)v$ term comes from the variation of the enthalpy as pressure varies from the standard pressure and $v = V/N_x$ is the specific volume per N_x entity. The standard pressure $(p°)$ and concentration $(C°)$ are introduced in the logarithms, such that the logarithmic contributions are zero at the standard conditions. Since $(p - p°)v < pv = k_B T$ is a relatively small contribution, we shall typically ignore it and just write

$$G^X_{(g)}(p) = G^X_{(g)}° + k_B T \cdot \ln\left(\frac{p}{p°}\right) \tag{3.13}$$

$$G^X_{(aq)}(C, p) = G^X_{(aq)}° + k_B T \cdot \ln\left(\frac{C}{C°}\right) \tag{3.14}$$

$$G^X_{(l)}(p) = G^X_{(l)}° \tag{3.15}$$

$$G^X_{(s)}(p) = G^X_{(s)} \tag{3.16}$$

These expressions are "idealized." Equations (3.13) and (3.14) assume that the atomic-scale entities are weakly (or non-)interacting, while Equations (3.14) to (3.16) assume the solvent, liquid, and solid to be incompressible. Real gas molecules and ions, of course, do interact and liquids are compressible and thus present variations from the formulas presented earlier. These variations are, however, typically rather limited and for the most part do not affect the discussion and conclusions in this text. We shall therefore throughout the text neglect deviations from ideal behavior.

We note that since we are considering the Gibbs free energy per atom, per molecule, or per reaction stoichiometry, this is by definition the chemical potential. We shall, however, generally refer to this quantity as the Gibbs free energy, in line with keeping the nomenclature for all other concepts, which we express in "per atomic-scale" entities.

We can use the expressions for the Gibbs energy (in conjunction with the requirement that reaction free energies are zero at equilibrium) to determine the equilibrium pressures and concentrations for given reactions. For a reaction A → B, the change in free energy is thus given by

$$\Delta G = \Delta G^{\circ} + k_{B}T\left(\ln\frac{p_{B}}{p^{\circ}} - \ln\frac{p_{A}}{p^{\circ}}\right) = \Delta G^{\circ} + k_{B}T\ln\frac{p_{B}}{p_{A}} \qquad (3.17)$$

At equilibrium, $\Delta G = 0$, and we thus obtain

$$\left.\frac{p_{B}}{p_{A}}\right|_{Eq} = e^{\frac{-\Delta G^{\circ}}{k_{B}T}} = K_{eq} \qquad (3.18)$$

where we have introduced the equilibrium constant, K_{eq}, of the reaction. This expression for the activities (pressures and concentrations of a reaction) as a function of the standard Gibbs energy of reaction is called the *law of mass action* for the reaction. In Appendix 3.1, we write up the general law of mass action for any reaction involving gases, solutes, liquids, and solids, based on the idealized Equations (3.25) to (3.28). This gives a general definition of the equilibrium constant for a reaction, as the unit-less quantity $K_{eq} = e^{-\Delta G^{\circ}/k_{B}T}$. We can use the generalized law of mass action to quickly write up equilibrium requirements. The equilibrium constant defined in this way is dimensionless because pressures and concentrations always occur (from Equations 3.13 and 3.14) in multiples of the standard pressure and the standard concentration. We may therefore just as well talk about pressures and concentrations in multiples of the standard pressure and concentration, thus dropping the p° and C° and thinking of pressures and concentrations as dimensionless quantities.

Consider as an example the ammonia synthesis reaction discussed in Chapter 2. The overall reaction is given by

$$N_{2} + 3H_{2} \rightarrow 2NH_{3} \qquad (3.19)$$

This should have a law of mass action of the form

$$K_{eq} = e^{-\frac{\Delta G^{\circ}}{k_{B}T}} = \left.\frac{p_{NH_{3}}^{2}}{p_{N_{2}}p_{H_{2}}^{3}}\right|_{Eq} \qquad (3.20)$$

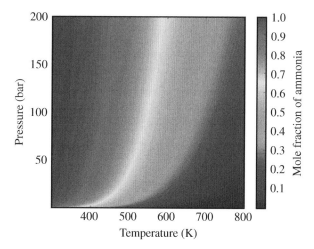

FIGURE 3.1 Equilibrium concentration of ammonia plotted as a function of temperature and pressure assuming a N_2:H_2 ratio of 1:3. Low temperatures and high pressures favor ammonia, whereas high temperatures lead to a shift in equilibrium toward the reactants. Industrially, ammonia synthesis is performed at approximately 700 K and at a total pressure on the order of 100 bar. (*See insert for color representation of the figure.*)

Ammonia synthesis is an exothermic reaction with an experimentally observed $\Delta H° = -0.95$ eV. As we will discuss in more detail in Chapter 7, ammonia synthesis is industrially carried out at fairly high temperatures (~700 K) in order to attain reasonably high turnover rates. Since we convert four gas-phase molecules into two and gas-phase molecules have a significant amount of entropy, this leads to a $\Delta S°$ of -2.05 meV/K. At 700 K, this gives $\Delta G° = 0.49$ eV and therefore a relatively low-equilibrium constant of $K_{eq} \sim 0.0002$. At these high temperatures, the equilibrium is thus shifted strongly back toward the reactants, N_2 and H_2. High pressures are therefore necessary in order to attain a reasonable conversion of reactants into products. This is shown in Figure 3.1 where the equilibrium concentration of ammonia is shown as a function of temperature and total pressure, assuming a stoichiometric N_2:H_2 ratio of 1:3. Note that if we were actually designing an industrial catalytic reactor, we would have to make a more accurate description of the free energetics, including a treatment of the nonideal properties of the gases, but assuming ideal behavior is sufficiently accurate for describing the qualitative features of the problem.

3.2 THE ADSORPTION ENTROPY

In order to determine the change in free energy, ΔG, of an adsorption process, we need to know both the adsorption energy and the adsorption entropy. The previous chapter dealt with the adsorption energy. We now turn to the gas-phase and adsorption-phase entropy. The total entropy, S_{tot}, of a molecule in the pure noninteracting gas-phase has four contributions:

$$S_{tot} = S_{trans} + S_{rot} + S_{vib} + S_{el} \qquad (3.21)$$

that is, the translational, rotational, vibrational, and electronic contributions, corresponding to the different quantum states of these independent degrees of freedom. If it is a monatomic gas, there will of course be no rotational and vibrational contributions to the entropy as the gas molecules in a monatomic gas do not have these internal degrees of freedom. We shall now establish some useful rules of thumb for thinking about the magnitude of the entropy of gas-phase atoms and molecules and the entropy that is lost once an atom or a molecule adsorbs on a surface.

The standard entropy of small gas-phase molecules, such as N_2 or CO, is of the order of 2 meV/K. Their entropic contribution to the Gibbs free energy at room temperature (and at 1 bar) is therefore approximately −0.6 eV/molecule and increases with temperature to about −1 eV at 500 K. Most smaller molecules involved in heterogeneous catalysis have entropies of this order of magnitude; one important exception is H_2 having a standard entropy of only 1.35 meV/K. Note that, for example, by using these approximate values, one will obtain $\Delta S° = -2.05$ meV/K for the ammonia synthesis reaction, which is precisely the experimentally observed value (see the preceding text). By far, the largest fraction of the standard entropic contributions arises from the translations, which result in an enormous number of (particle-in-a-box type) energy states. The vibrational and rotational parts constitute a minor fraction of the gas-phase entropy. There are typically very few accessible electronic states, resulting in a very small electronic entropy contribution, and we shall generally ignore this contribution. The nonstandard contributions (the pressure dependence) to the gas-phase entropy come from the translational modes.

When a molecule adsorbs on the surface, it will lose a major part of its gas-phase entropy, as it loses the translational freedom from the gas phase. The translational and rotational degrees of freedom typically become constrained and turn into vibrational modes (at least at low temperatures—at higher temperatures, they might become frustrated translational or frustrated rotational). The total contribution of a vibrational mode with frequency, ν_i, to the standard Gibbs free energy is (except for the zero-point energy (ZPE) contributions that are discussed in Chapter 2 and 4)

$$G°_{\nu_i} = k_B T \ln\left(1 - e^{\frac{-h\nu_i}{k_B T}}\right) \qquad (3.22)$$

We shall therefore typically think of $S°_{\nu_i}$ as being

$$S°_{\nu_i} = -k_B \ln\left(1 - e^{\frac{-h\nu_i}{k_B T}}\right) \qquad (3.23)$$

The full entropic term is actually more complex than this; there is an additional term, but this other term exactly cancels the temperature dependence (the vibrational C_p term) of the enthalpy. As can be seen from Equation (3.23), $S°_{\nu_i}$ depends on the

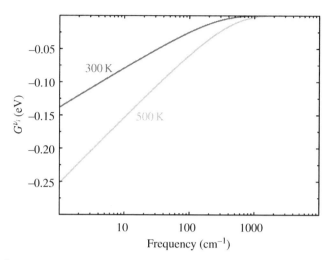

FIGURE 3.2 $G^\circ_{v_i}(T)$ plotted as a function of vibrational frequency for 300 and 500 K (ignoring the ZPE contribution).

frequency of vibration of the adsorbed molecule. Large frequencies will only give small values of $S^\circ_{v_i}$, while small frequencies lead to larger values of $S^\circ_{v_i}$. To give an estimate of the magnitude of $S^\circ_{v_i}$, we show in Figure 3.2 how $G^\circ_{v_i}(T)$ for a single vibrational mode varies as a function of the frequency due to the vibrational entropy (ignoring the ZPE contribution). It can be seen that only frequencies that are around 50 cm^{-1} or lower give significant contributions to $G^\circ_{v_i}(T)$. Remembering that gas-phase molecules have an entropic contribution to the standard Gibbs energy of approximately -1 eV at 500 K, we can see that unless there are a number of very-low-frequency vibrational modes of the adsorbed molecule on the surface, it can be assumed as a rough approximation that the molecule loses all of its entropy upon adsorption:

$$\Delta S^\circ_{ads} = S^\circ_{ads} - S^\circ_{gas\text{-}phase} \approx -S^\circ_{gas\text{-}phase} \tag{3.24}$$

In addition to the contribution to the adsorption entropy from the vibrational (frustrated translational and rotational) degrees of freedom, there is an entropy contribution from the different configurations the adsorbate can have on the surface. Consider the adsorption of a gas-phase molecule onto a surface:

$$A_{(g)} + * \rightarrow A*$$

The asterisk denotes a surface site where A can adsorb to form the adsorbed species, A*. The different ways in which N adsorbates can occupy N_0 surface sites give rise to a configurational entropy contribution that depends on the coverage $\theta = N/N_0$:

$$S^{conf}_{ads} = -k_B \ln\left(\frac{\theta_A}{\theta_*}\right) \tag{3.25}$$

For a derivation of this expression, see Appendices 3.2 and 3.3. It is seen that the coverage plays the role of an activity like the gas-phase pressure or concentration of species in solution. This expression makes it natural to choose the "standard" for coverage as the point where the coverage of the adsorbate in question is equal to the coverage of free sites. When only one species is present on the surface, the standard coverage thus corresponds to a surface that is half covered by the adsorbate. This looks *a priori* quite similar to the logarithmic contribution to the nonstandard term in the entropy of a gas or a solution. It is, however, different in the sense that the coverage of free sites, θ_*, is directly linked to the coverage of the adsorbate through the site balance equation (which states that the sum of fractional coverages is one).

3.3 ADSORPTION EQUILIBRIA: ADSORPTION ISOTHERMS

We will now examine how the equilibrium surface coverages depend on the adsorption strength of the adsorbing molecule and the influence of temperature and pressure. The equilibrium coverage is given by the requirement that the Gibbs free energy of adsorption is zero:

$$\Delta G = G_{ads} - G_{gas} = 0 \tag{3.26}$$

This means that

$$\Delta G = \Delta G^\circ - T\left(S_{ads}^{conf} - \left(-k_B \ln\left(\frac{p_A}{p^\circ}\right)\right)\right) = 0 \tag{3.27}$$

which gives us the corresponding law of mass balance:

$$K_A = e^{-\frac{\Delta G^\circ}{k_B T}} = \left(\frac{\theta_A}{\theta_*}\right)\left(\frac{p_A}{p^\circ}\right)^{-1} \tag{3.28}$$

or

$$\frac{\theta_A}{\theta_*} = K_A\left(\frac{p_A}{p^\circ}\right) \tag{3.29}$$

Remembering that since pressures typically enter in multiples of the standard pressure, p°, like here, we usually drop the p°, and let p_A describe a dimensionless pressure. We thus write

$$\theta_A = K_A p_A \theta_* \tag{3.30}$$

where the fractional coverage, the equilibrium constant, as well as the pressure are all dimensionless. We have here also implicitly used the convention that the number

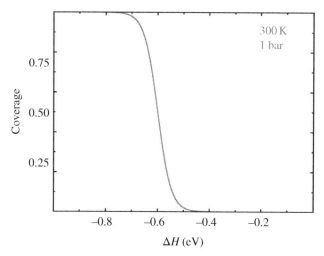

FIGURE 3.3   Coverage of species A plotted as a function of its adsorption strength, ΔH. Conditions are $300\,K$ and $1\,bar$ of A. The loss of entropy upon adsorption is $-0.002\,eV$.

of sites available on a surface is normalized to unity, such that the sum of all sites totals 1:

$$\theta_* + \theta_A = 1 \tag{3.31}$$

Substituting Equation (3.30) into Equation (3.31) gives

$$\left(1 + K_A p_A\right)\theta_* = 1 \tag{3.32}$$

and hence

$$\theta_* = \frac{1}{1 + K_A p_A} \tag{3.33}$$

The coverage of A is thus

$$\theta_A = \frac{K_A p_A}{1 + K_A p_A} \tag{3.34}$$

This expression for the equilibrium coverage of an adsorbate as a function of the gas-phase pressure and the standard reaction Gibbs free energy is called a *Langmuir isotherm*.

Figure 3.3 shows how θ_A changes with ΔH, that is, the adsorption strength of molecule A on a particular surface. We assume here an entropy loss of molecule A upon adsorption of $-0.002\,eV/K$. It can be seen that the coverage of A on the surface depends strongly on its adsorption strength. Since the $-T\Delta S°$ term in the Gibbs

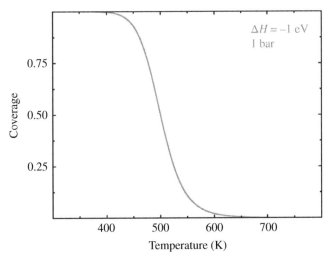

FIGURE 3.4 Coverage of species A plotted as a function of temperature for an adsorbate with adsorption strength, ΔH, of −1 eV, a gas-phase pressure of 1 bar, and a standard adsorption entropy of −2 meV/K.

energy is approximately 0.6 eV at 300 K, the equilibrium constant becomes on the order of 1 at $\Delta H = -0.6$ eV, and at this adsorption strength, we find $\theta_A = 0.5$. Note the strong dependence on ΔH as the surface sites shift from occupied to free following the Langmuir isotherm; there is a relatively narrow window of roughly 0.2 eV (at 300 K) within which θ_A shifts from essentially 1 to almost 0.

The coverage θ_A also depends strongly on the temperature of the system via the entropy term in the standard adsorption Gibbs energy in the equilibrium constant. A shift in temperature will lead to a strong change in the coverage of A. Using again the rule of thumb that most (smaller) gas-phase molecules have entropies on the order of 2 meV/K, Figure 3.4 shows the coverage of A, θ_A, as a function of temperature for an adsorption strength, ΔH, of −1 eV. An increase in temperature leads to an increase in $-T\Delta S$ and will hence decrease the adsorption equilibrium constant and the coverage of A. The strong dependence on the temperature becomes clear as the coverage changes from 0 to 1 due to a change in the temperature of less than 200 K.

Lastly, since θ_A also depends on the pressure p_A, we will show how pressure influences θ_A. Figure 3.5 shows the coverage of A plotted as a function of the pressure of A for three different adsorption strengths. When ΔH is very negative, θ_A will be large even at low pressures of A. At less negative values of ΔH (where binding to the surface is less favorable), one can significantly increase the θ_A with an increase in pressure. The pressure dependence is, however, less pronounced than the dependence of θ_A on ΔH and T.

In most catalytic processes, two or more molecules react with each other. The coverages of these different surface species depend on each other, since the adsorbates compete for the same adsorption sites. We will show this dependence for an example

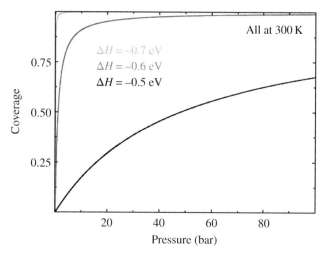

FIGURE 3.5 Coverage of species A plotted as a function of pressure at three different values of adsorption enthalpy, ΔH. The temperature is 300 K and the standard adsorption entropy is $-2\,\text{meV}$. (*See insert for color representation of the figure.*)

where we have species A and B adsorbing on the surface. As shown in Equation (3.30), the coverages of species A and, similarly, species B can be written as

$$\theta_A = K_A p_A \theta_* \tag{3.35}$$

$$\theta_B = K_B p_B \theta_* \tag{3.36}$$

If we constrain the number of free sites to one (as we shall typically always do), we get

$$\theta_* + \theta_A + \theta_B = 1 \tag{3.37}$$

This gives

$$\theta_* = \frac{1}{1 + K_A p_A + K_B p_B} \tag{3.38}$$

and hence

$$\theta_A = \frac{K_A p_A}{1 + K_A p_A + K_B p_B} \tag{3.39}$$

$$\theta_B = \frac{K_B p_B}{1 + K_A p_A + K_B p_B} \tag{3.40}$$

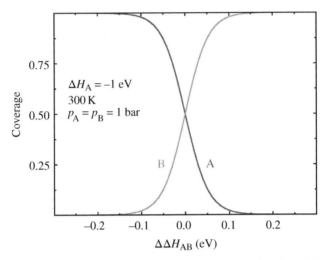

FIGURE 3.6   Coverages of species A and species B plotted as a function of their difference in ΔH, $\Delta\Delta H_{AB}$. $\Delta\Delta H_{AB}$ is defined as $\Delta H_A - \Delta H_B$, where ΔH_A is kept at a constant value of -1 eV. The temperature is kept at 300 K and the pressures of A and B are 1 bar. The standard entropy of adsorption is assumed to be -2 meV.

Differences in adsorption energies of even as little as 0.1 eV lead to a total dominance of the stronger adsorbing species on the surface. This is observed in Figure 3.6 where the coverages of A and B are plotted as a function of their difference in adsorption strength $\Delta\Delta H_{AB} = \Delta H_A - \Delta H_B$ for an adsorption enthalpy of species A of $\Delta H_A = -1$ eV. At $\Delta\Delta H_{AB} = 0$, both species have a surface coverage close to 0.5. Already when A binds only 0.1 eV stronger than B, the coverage of A totally dominates the surface (at this temperature). There is thus only a narrow range of adsorption energy differences in which the two species are coadsorbed on the surface in appreciable amounts. As soon as one species binds about 0.1 eV stronger than all other species, it will totally dominate the surface coverage at room temperature (and if it binds just 0.2 eV stronger than all other species, it will totally dominate the surface coverage at 2 times the room temperature ~600 K).

CO POISONING OF A PEM FUEL CELL

Pt is used as a catalyst in fuel cell reactors where it catalyzes the oxidation of H_2 to protons (Gasteiger et al., 1994). One of the factors determining the activity of the catalyst is the surface area available to catalyze this process (i.e., the number of sites where H_2 can adsorb). The main source of H_2 is currently natural gas, from which H_2 is produced via steam reforming and water–gas shift. This process usually results in H_2 gas that is contaminated with small amounts of CO. Since CO binds strongly to the Pt surface, CO poisoning is an important issue for hydrogen fuel cell catalysts. This is an example of competitive adsorption—CO versus H_2. We can estimate the severity of CO poisoning on Pt(111) using data from the CatApp (see Chapter 2). We get a CO adsorption energy of $-1.2\,\text{eV}$ and a dissociative H_2 adsorption energy of $-0.7\,\text{eV}$ on the Pt(111) surface. We showed earlier that a difference in adsorption energies of $0.1\,\text{eV}$ leads to a total dominance of one species. We will now try to estimate how low of a CO pressure is needed in order to avoid poisoning the surface. H_2 adsorbs dissociatively on the Pt surface:

$$H_2 + 2* \rightarrow 2H* \tag{3.41}$$

The configurational entropy of adsorbed H is therefore

$$S_H^{conf} = -k_B \ln\left(\frac{\theta_H^2}{\theta_*^2}\right) \tag{3.42}$$

This gives the following expressions for the coverages of H and CO:

$$\theta_H = \frac{\sqrt{K_H p_{H_2}}}{1 + \sqrt{K_H p_{H_2}} + K_{CO} p_{CO}} \tag{3.43}$$

$$\theta_{CO} = \frac{K_{CO} p_{CO}}{1 + \sqrt{K_H p_{H_2}} + K_{CO} p_{CO}} \tag{3.44}$$

Figure 3.7 shows how even ppm amounts of CO severely reduce the H coverage on the surface and hence the ability of Pt to act as a hydrogen oxidation catalyst. This explains the observed CO poisoning.

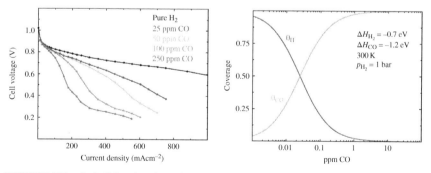

FIGURE 3.7 *Left*: CO poisoning of a PEM fuel cell. Adapted from Oetjen et al. (1996). *Right*: competitive adsorption between CO and H_2 on the Pt(111) surface as function of CO pressure. Conditions are 300 K and 1 bar of H_2.

3.4   FREE ENERGY DIAGRAMS FOR SURFACE CHEMICAL REACTIONS

In Chapter 2, it was shown how to construct potential energy diagrams (PEDs) from adsorption and reaction energies. These energies can, for example, be obtained from the CatApp (see Fig. 2.4), and Figure 2.5 shows the PED of ammonia synthesis on the stepped Ru(0001) surface. Ammonia synthesis (Equation 3.19) is an exothermic reaction where the standard entropy of the products is lower than that of the reactants (4 gas-phase molecules are converted to 2). We already showed in Figure 3.2 that high pressures are needed to shift the equilibrium to favor ammonia over nitrogen and hydrogen at higher temperatures. While PEDs are good tools to obtain a simple overview of the elementary reaction steps in a process, they often fail to give insight into reactions that involve large shifts in entropy (such as ammonia synthesis) at a level sufficient to make even qualitative predictions about relative reaction rates. We shall therefore introduce the *Gibbs free energy diagram*. These diagrams take the effects of temperature and pressure into account by showing the "configurationally corrected Gibbs free energy levels," ΔG_A^{cc}, of every reaction step, A, with respect to the reactants:

$$\Delta G_A^{cc} = \Delta H_A - T\left(S_A{}^\circ - S_R\right) \tag{3.45}$$

where ΔG_A^{cc} is understood to be the full Gibbs reaction energy from the reactants but in which the entropy of all adsorbates are calculated in their "standard states" given by $\theta_i = \theta_*$.

The *Gibbs free energy diagram* for ammonia synthesis is shown in Figure 3.8. The diagram was derived from the energies shown in Figure 2.6, with the values being ZPE corrected. The gas-phase entropy is included, while entropy contributions of adsorbed species are assumed to be zero.

Figure 3.8 shows the free energy diagram for ammonia synthesis at different temperatures. The effect of temperature is rather strong for ammonia synthesis where 4 molecules are converted into 2. Ammonia synthesis is exothermic, but high temperatures act against that by increasing the $T\Delta S$ term. While ammonia synthesis is still exergonic at 300 K, it becomes about neutral at 500 K and is uphill by almost 0.5 eV at 700 K. Since we neglect entropic contributions from adsorbed species, the effect of temperature can only be seen in adsorption and desorption processes.

Figure 3.8 shows why high temperature is needed for the ammonia synthesis to proceed rapidly. At 300 K, the adsorbed N-containing species are so stable with respect to NH_3 in the gas phase that the surface will be completely covered as soon as a small amount of ammonia is formed. This means that there is no place for N_2 to dissociate and the rate is extremely low. Only at temperatures around 700 K does the "thermodynamic sink" associated with adsorbed NH_x disappear. At this high temperature, however, the reaction is uphill in free energy (endergonic). In order to push the equilibrium toward some ammonia conversion, the pressures of N_2 and H_2 need to be high (see also Figure 3.1). We will discuss the influence of pressure in the following.

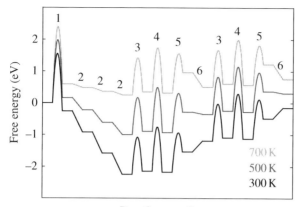

FIGURE 3.8 Gibbs free energy diagram for ammonia synthesis on the stepped Ru(0001) surface at 300, 500, and 700 K. The numbers correspond to the six different reaction steps: (1) $N_2 + 2^* \rightarrow 2N^*$; (2) $H_2 + 2^* \rightarrow 2H^*$; (3) $N^* + H^* \rightarrow NH^* + ^*$; (4) $NH^* + H^* \rightarrow NH_2^* + ^*$; (5) $NH_2^* + H^* \rightarrow NH_3^* + ^*$; (6) $NH_3^* \rightarrow NH_3 + ^*$ (see also Fig. 2.6). Data was obtained from CatApp and corrected for ZPE and entropy contributions. The entropy of adsorbed species was assumed to be zero. A pressure of 1 bar was assumed for all species involved.

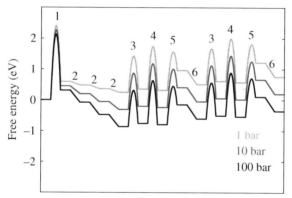

FIGURE 3.9 Gibbs free energy diagram for ammonia synthesis on the stepped Ru(0001) surface at 1, 10, and 100 bar. The numbers correspond to the six different reaction steps (see also Figs. 2.6 and 3.9) as defined in the preceding text. Data was obtained from the CatApp and corrected for ZPE and entropy contributions. The entropy of adsorbed species was assumed to be zero. Reaction conditions are as follows: $T = 700$ K, $N_2:H_2 = 1:3$, conversion to ammonia $= 10\%$.

If the pressure term is included for varying pressures as described in Equation (3.17), a quantitative picture of the pressure dependence of the different reaction steps is achieved. Figure 3.9 shows the free energy diagram of ammonia synthesis at 700 K and 10% conversion toward ammonia. At ambient pressures (1 bar total pressure), the reaction is endergonic by about 0.3 eV (note that here the *total* pressure of all reactants and products is 1 bar, whereas neglect of the pressure term assumes 1 bar for *each* of the gas-phase molecules involved). A 10% conversion cannot be therefore achieved at 700 K and 1 bar; the reaction would simply run backward until equilibrium is obtained. (The threshold for conversion as a function of temperature and pressure is also shown in Fig. 3.2.) A pressure of 10 bar increases the adsorption of gas-phase species and leads to an approximately thermoneutral Gibbs energy of reaction. The conversion can be further improved when going to even higher pressures (here shown for 100 bar) where the reaction is now downhill in free energy.

APPENDIX 3.1   THE LAW OF MASS ACTION AND THE EQUILIBRIUM CONSTANT

Suppose we have a *very* general reaction involving a number (N_g^R) of gaseous reactants (X_i^R) with stoichiometric coefficients, $\kappa_{g_i}^R$. The reaction also involves a number (N_{aq}^R) of solvated reactants (A_j^R) with stoichiometric coefficients, $\kappa_{aq_j}^R$. The reaction furthermore involves similarly defined liquid $\left(L_k^R\right)$ and solvent $\left(S_h^R\right)$ reactants, with appropriate stoichiometric coefficients. These react to form a number (N_g^P) of gaseous products (X_i^P) with stoichiometric coefficients, $\kappa_{g_i}^P$; a number (N_{aq}^P) of solvated products (A_j^P) with stoichiometric coefficients, $\kappa_{aq_j}^P$; and so on. The products can also react in the backward direction (as all elementary reactions can):

$$\sum_{i=1}^{N_g^R}\kappa_{g_i}^R X_i^R + \sum_{j=1}^{N_{aq}^R}\kappa_{aq_j}^R A_j^R + \sum_{k=1}^{N_l^R}\kappa_{l_k}^R L_k^R + \sum_{h=1}^{N_s^R}\kappa_{s_h}^R S_h^R \rightleftarrows$$

$$\sum_{i=1}^{N_g^P}\kappa_{g_i}^P X_i^P + \sum_{j=1}^{N_{aq}^P}\kappa_{aq_j}^P A_j^P + \sum_{k=1}^{N_l^P}\kappa_{l_k}^P L_k^P + \sum_{h=1}^{N_s^P}\kappa_{s_h}^P S_h^P$$

(A.3.1.1)

Equilibrium for this reaction occurs when the Gibbs energy of reaction is zero. We thus calculate the Gibbs energy of reaction (per stoichiometric reaction)

$$\Delta G = \sum_{i=1}^{N_g^P}\kappa_{g_i}^P G_{X_i^P} + \sum_{j=1}^{N_{aq}^P}\kappa_{aq_j}^P G_{A_j^P} + \sum_{k=1}^{N_l^P}\kappa_{l_k}^P G_{L_k^P} + \sum_{h=1}^{N_s^P}\kappa_{s_h}^P G_{S_h^P}$$

$$- \sum_{i=1}^{N_g^R}\kappa_{g_i}^R G_{X_i^R} - \sum_{j=1}^{N_{aq}^R}\kappa_{aq_j}^R G_{A_j^R} - \sum_{k=1}^{N_l^R}\kappa_{l_k}^R G_{L_k^R} - \sum_{h=1}^{N_s^R}\kappa_{s_h}^R G_{S_h^R}$$

(A.3.1.2)

and reduce out the standard contributions in the term $\Delta G°$ using the expressions for ideal gases, solutions, liquids, and solids in Equations (3.25) to (3.28):

$$\Delta G = \Delta G^\circ + k_B T \sum_{i=1}^{N_g^P} \kappa_{g_i}^P \ln\left(\frac{P_{X_i^P}}{p^\circ}\right) + k_B T \sum_{j=1}^{N_{aq}^P} \kappa_{aq_j}^P \ln\left(\frac{C_{A_j^P}}{C^\circ}\right)$$

$$- k_B T \sum_{i=1}^{N_g^R} \kappa_{g_i}^R \ln\left(\frac{P_{X_i^R}}{p^\circ}\right) - k_B T \sum_{j=1}^{N_{aq}^R} \kappa_{aq_j}^R \ln\left(\frac{C_{A_j^R}}{C^\circ}\right)$$

(A.3.1.3)

In equilibrium, we thus have

$$-\frac{\Delta G^\circ}{k_B T} = \sum_{i=1}^{N_g^P} \kappa_{g_i}^P \ln\left(\frac{P_{X_i^P}}{p^\circ}\right) + \sum_{j=1}^{N_{aq}^P} \kappa_{aq_j}^P \ln\left(\frac{C_{A_j^P}}{C^\circ}\right) - \sum_{i=1}^{N_g^R} \kappa_{g_i}^R \ln\left(\frac{P_{X_i^R}}{p^\circ}\right) - \sum_{j=1}^{N_{aq}^R} \kappa_{aq_j}^R \ln\left(\frac{C_{A_j^R}}{C^\circ}\right)$$

(A.3.1.4)

Exponentiation of both sides of this equation results in an expression that is called the *law of mass action* (due to the assumptions made earlier that it is not really a *law*, but rather a rule of thumb, which holds well for ideal gases, ideal solutions, and incompressible solutions, liquids, and solids):

$$e^{-\frac{\Delta G^\circ}{k_B T}} = \left. \frac{\prod_{i=1}^{N_g^P}\left(\frac{P_{X_i^P}}{p^\circ}\right)^{\kappa_{g_i}^P} \prod_{j=1}^{N_{aq}^P}\left(\frac{C_{A_j^P}}{C^\circ}\right)^{\kappa_{aq_j}^P}}{\prod_{i=1}^{N_g^R}\left(\frac{P_{X_i^R}}{p^\circ}\right)^{\kappa_{g_i}^R} \prod_{j=1}^{N_{aq}^R}\left(\frac{C_{A_j^R}}{C^\circ}\right)^{\kappa_{aq_j}^R}} \right|_{Eq}$$

(A.3.1.5)

The value of the law of mass action is that it gives us a tool for determining equilibrium pressures and concentrations for a reaction for which we can calculate the standard free energy of reaction. Since the activities of the solids and liquids do not appear in the expression, one always has to check if so much of any of these has reacted that it is not present as a liquid or a solid state any longer. If this is the case, then the reaction described by ΔG° has run to completion, and the equilibrium has not been established. The dimensionless quantity, $e^{-(\Delta G^\circ / k_B T)}$, is a constant that the right-hand-side function of pressures and concentrations in the law of mass action (A.3.1.5) should satisfy at equilibrium, and we shall therefore call it the *equilibrium constant* for the reaction

$$K_{eq} = e^{-\frac{\Delta G^\circ}{k_B T}}$$

(A.3.1.6)

APPENDIX 3.2   COUNTING THE NUMBER OF ADSORBATE CONFIGURATIONS

The number of ways one can distribute N_A adsorbates on a clean surface with N_S surface sites can be calculated in the following way. For the first adsorbate, there are N_S available free sites to choose between; for the second adsorbate, there are $(N_S - 1)$ sites remaining to choose between; and so on. This process continues until the N_Ath adsorbate, for which there are $(N_S - N_A + 1)$ free sites left to choose between. The number of ways to place the N_A adsorbates is then product of the individual possibilities:

$$N_S \cdot (N_S - 1) \cdot (N_S - 2) \cdots (N_S - N_A + 1) = \frac{N_S!}{(N_S - N_A)!} \qquad \text{(A.3.2.1)}$$

However, the N_A adsorbates we have placed on the surface are indistinguishable. This means that if we exchange an adsorbate that was put in site "X" with an adsorbate that was put in site "Y," then the resulting configuration would be indistinguishable from the original state. Therefore, Equation (A.3.2.1) overcounts the number of physically distinguishable distributions by a factor equal to the number of ways we can order the N_A adsorbates (in the N_A chosen sites on the surface). The number of such ways can be described as the number of ways we can pick an adsorbate for site one, multiplied by the number of ways we can pick one for site two, etc., until all N_A adsorbates have been placed in their sites. For the first pick, there are N_A adsorbates to choose between; for the second choice, there are $(N_A - 1)$ to choose between; etc. So in total, there are $N_A!$ ways to order the adsorbates, and the total number of configurations, N_{conf}, is therefore given by

$$N_{conf}(N_S, N_A) = \frac{N_S!}{(N_S - N_A)! \cdot N_A!} \qquad \text{(A.3.2.2)}$$

This expression is often referred to as a binomial coefficient since it is the same as the coefficient in front of the x^{N_A}th term in the polynomial expansion of the binomial $(1 + x)^{N_S}$.

APPENDIX 3.3   CONFIGURATIONAL ENTROPY OF ADSORBATES

The entropy stemming from a number of different possible states with the same energy is $S = k_B \ln \Omega$, where Ω is the number of distinguishable microscopic states. This is the Boltzmann formula for the entropy of a closed system. (This is a particular case of the more general Gibbs equation for entropy defined for open (as well as closed) systems: $S = -k_B \sum_i p_i \ln p_i$, where p_i is the probability of finding the system in the microscopic state i.) The total configurational entropy of a system of N_A adsorbates distributed in N_S random sites on a surface is therefore

$$S_{\text{conf}}^{\text{system}} = k_{\text{B}} \cdot \ln \frac{N_{\text{S}}!}{\left(N_{\text{S}} - N_{\text{A}}\right)! \cdot N_{\text{A}}!} \tag{A.3.3.1}$$

Entropy generally is expressed as a product of two factors: one that is proportional to the system size and another that is independent of the system size. To obtain a more generally useful expression than Equation (A.3.3.1), we employ an approximation to the factorial expression (retaining only terms of at least order n):

Stirling's approximation (to order n)—$\ln n! \approx n \cdot \ln n - n$

By which, it is obtained that

$$S_{\text{conf}}^{\text{system}} = k_{\text{B}} \cdot \left(N_{\text{S}} \cdot \ln N_{\text{S}} - \left(N_{\text{S}} - N_{\text{A}}\right) \cdot \ln\left(N_{\text{S}} - N_{\text{A}}\right) - N_{\text{A}} \cdot \ln N_{\text{A}}\right) \tag{A.3.3.2}$$

$$S_{\text{conf}}^{\text{system}} = -N_{\text{S}} \cdot k_{\text{B}} \cdot \left(\frac{\left(N_{\text{S}} - N_{\text{A}}\right)}{N_{\text{S}}} \cdot \ln \frac{\left(N_{\text{S}} - N_{\text{A}}\right)}{N_{\text{S}}} + \frac{N_{\text{A}}}{N_{\text{S}}} \cdot \ln \frac{N_{\text{A}}}{N_{\text{S}}}\right) \tag{A.3.3.3}$$

Now, using that $\theta_{\text{A}} = N_{\text{A}}/N_{\text{S}}$ and $\theta_* = (N_{\text{S}} - N_{\text{A}})/N_{\text{S}}$, we get the expression

$$S_{\text{conf}}^{\text{system}} = -N_{\text{S}} \cdot k_{\text{B}} \cdot \left(\left(1 - \theta_{\text{A}}\right) \cdot \ln\left(1 - \theta_{\text{A}}\right) + \theta_{\text{A}} \cdot \ln \theta_{\text{A}}\right) \tag{A.3.3.4}$$

This expresses the *total entropy* for a system with N_{S} surface sites. Throughout the book, we take an "atomic-scale point of view," so we typically think of the thermodynamic quantities on a per atom, per molecule, or per reaction basis. We thus obtain the differential configurational adsorption entropy (the configurational entropy gained by the system through the adsorption of one extra adsorbate) as a function of coverage, by differentiating the total system entropy per surface site with respect to the coverage:

$$S_{\text{conf}}^{\text{ads}} = \frac{\partial\left(S_{\text{conf}}^{\text{system}} / N_{\text{S}}\right)}{\partial \theta_{\text{A}}} = -k_{\text{B}} \ln\left(\frac{\theta_{\text{A}}}{\theta_*}\right) \tag{A.3.3.5}$$

The configurational entropy gained upon adsorption thus goes to infinity when the coverage goes to zero (and goes to negative infinity as the coverage goes to 1). This is the fundamental reason why it is essentially impossible to make a totally clean or totally covered surface.

REFERENCES

Gasteiger HA, Markovic N, Ross PN, Cairns EJ. Carbon monoxide electrooxidation on well-characterized platinum-ruthenium alloys. J Phys Chem 1994;98:617.

Oetjen HF, Schmidt VM, Stimming U, Trila F. Performance data of a proton exchange membrane fuel cell using H_2/CO as fuel gas. J Electrochem Soc 1996;143:3838.

FURTHER READING

Kittel C, Kroemer H. *Thermal Physics*. 2nd ed. San Francisco: W.H. Freeman; 1980.

Schlögl R. Catalytic synthesis of ammonia—A "never-ending story"? Angew Chem Int Ed 2003;42:2004.

Silbey RJ, Alberty RA, Bawendi MG. *Physical Chemistry*. 4th ed. John Wiley & Sons, Inc: Hoboken; 2005.

4

RATE CONSTANTS

As discussed in the previous chapters, much can be learned about the catalytic properties of a surface from analyzing potential energy diagrams and free energy diagrams for the process intermediates reacting over the surface. As we saw in Chapter 3, the free energy calculations/diagrams let us determine the equilibria for elementary and nonelementary reactions. Through knowledge of the adsorption free energies and the dependence of the configurational adsorption entropy on coverages, they also let us determine the coverages of various species on the surface under equilibrium conditions.

Equilibrium considerations can certainly be useful, but catalysis is in the end really all about speeding up a reaction, where the reactants and products are not in equilibrium, so we must be able to address reaction rates to obtain a detailed picture of what goes on at catalytic surfaces under relevant conditions. We did not specify exactly how to get to the elementary rates from knowledge of the Gibbs free energy diagrams, but have so far only postulated an Arrhenius-type expression for rate constants in terms of (an as-of-yet undetermined) prefactor and an activation energy that we could obtain already from the potential energy surface (PES). We have also not yet addressed how to determine coverages of adsorbates when there are reactions occurring at the surface out of equilibrium. In order to determine the out-of-equilibrium coverages during a catalytic reaction, we will need to quantitatively model the kinetics, and this is the topic of Chapter 5. In order to quantitatively model the kinetics in Chapter 5, we will need to specify rate constants for elementary reactions.

Fundamental Concepts in Heterogeneous Catalysis, First Edition. Jens K. Nørskov,
Felix Studt, Frank Abild-Pedersen and Thomas Bligaard.
© 2014 John Wiley & Sons, Inc. Published 2014 by John Wiley & Sons, Inc.

This chapter, which will at times become rather technically involved and perhaps appear only remotely connected to catalysis, aims to specify exactly what is meant by the term "rate constant" and how it relates to the Gibbs free energy surface. The reason we have included it anyhow in a book on the "fundamental concepts in catalysis" is that it strengthens the foundation under the free energy diagrams as a central concept and tool for analyzing catalytic reactions on surfaces. It will also be putting the concepts of activation energy, activation entropy, and transition state (TS) zero-point energy corrections on a firmer foundation. Since this chapter may be significantly more useful for the reader who has the ambition of simulating catalytic reaction rates than to other readers, some might want to just accept the central result of the chapter and move on to Chapter 5. The central result we shall discuss is that the rate constant for an elementary reaction is given by

$$k = \frac{k_B T}{h} e^{-\Delta G_{TS}°/k_B T} \tag{4.1}$$

where $\Delta G_{TS}° = \Delta E_{TS}° - T \Delta S_{TS}° = E_a - T \Delta S_{TS}°$ is the standard Gibbs free energy in a so-called TS minus the standard Gibbs free energy in the reactant state. This implies that we have a practical way to calculate the prefactor, $v = \frac{k_B T}{h} e^{\Delta S_{TS}°/k_B}$, in the Arrhenius expression (Equation 2.1).

4.1 THE TIMESCALE PROBLEM IN SIMULATING RARE EVENTS

Having established a PES, one can (at least in principle) calculate reaction rates directly, by numerically integrating the dynamics of the atoms for "long enough" periods of time to obtain reliable statistics on the reaction rate. It turns out, however, that this is rarely a practical approach. Classical molecular dynamics integration entails starting the system with a kinetic energy corresponding to the temperature of the system, in a reasonable position or structure located in a region of the PES, which we would classify as the "reactant region." Then we would develop the position of the system according to Newtonian mechanics, by calculating the force on the system (as the negative derivative of the PES). We simultaneously need to ensure that statistical fluctuations representing random thermal fluctuations (stemming from the system's thermal contact with a heat reservoir) enter in a reasonable way. The rate could then in principle be measured as the inverse time until the system reaches the region of the PES, which we would identify as belonging to the "product region" of the elementary reaction. Perhaps we would need to redo this a number of times in order to obtain reasonable statistics. The main problem with this direct dynamics integration approach for catalysis applications is that the important catalytic reactions are those that are slow, since these reactions limit the overall catalytic rate. In order to reliably integrate Newton's 2nd law, one needs to take time steps of a size that is on the order of inverse molecular vibration frequencies. As we shall see in this chapter, this corresponds to taking on the order of 10^{13} time steps in the molecular dynamics integration for modeling a typical catalytic rate of 1 s^{-1}. This is perhaps 6 orders of magnitude more time steps than what is typically affordable on a modern computer and requires very detailed information of the PES, which is typically costly to obtain.

Fortunately, a number of methods exist to treat the problem of modeling rare events, and reaction rates can often be obtained without carrying out full molecular dynamics simulations. Most reaction rate theories for elementary processes build upon the ideas introduced in transition state theory (TST). This theory establishes how to get from knowledge of the PES to simple estimates of the rate, how the activation energy and prefactor in the Arrhenius expression come about, and what its building blocks are. We shall follow the simplest path to deriving TST, by starting from a classical point of view and then substituting in the correct quantum statistical mechanics elements when necessary. The reason for performing such a slightly messy two-step derivation is twofold. First of all, it is significantly more transparent than a thorough quantum statistical mechanical treatment, and secondly, the true quantum contributions to the rate (quantum tunneling) end up being extremely involved to treat and are typically important only for lower-temperature reactions than those relevant for catalytic applications. Having obtained TST, we shall also discuss how to go beyond TST to calculate exact "classical" rates, as well as how to make the so-called harmonic approximation to TST (harmonic transition state theory (HTST)), which in many cases is adequately reliable for treating elementary processes at surfaces.

4.2   TRANSITION STATE THEORY

TST approaches the problem of calculating a reaction rate for a rarely occurring reactive event by separating the PES into two regions, one being the reactant region (R) and the other being the product region (P). The reactant region defines the general region in which the system can be found before reacting, and the product region defines what is thought of as a product of the elementary reaction in question. The border between the two regions, a so-called separatrix, is the TS. We shall assume that we can divide up the whole configuration space of the system such that any configuration belongs to either the reactant region, the product region, or the TS. The lowest energy configurations in the reactant and product regions are often referred to as the initial state (IS) and final state (FS), respectively. The lowest energy point in the TS is often also itself confusingly referred to as "the TS," especially if this is a first-order saddle point (a stationary point with one mode of negative curvature) on the PES; in which case, it is often used for making a harmonic approximation to TST, as we shall discuss. We shall here restrict the term TS to refer to the separatrix between the reactant and product regions. If entropy effects were to be neglected, the natural choice would be to let the TS follow the energy ridge between the reactant and product regions (as drawn in Fig. 4.1).

It may look easier than it typically is to determine a "good" TS on the PES. Since the TS has zero thickness along the reaction path, it is an object of dimensionality one

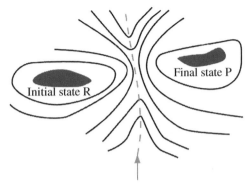

D-1 dimensional dividing surface, TS

FIGURE 4.1 Separation of the configuration space used in transition state theory into three regions: the reactant region, the product region, and the transition state, which is a dividing surface that separates the reactant and product regions.

lower than the full dimensionality of the PES. In Figure 4.1, for example, the TS is the dashed red line of dimensionality 1 on a PES of dimensionality 2. For a system of N_{atoms} atoms, the dimensionality of the PES is $3N_{atoms}$, since each atom has 3 spatial degrees of freedom (in the absence of any constraints). Already for systems with even a relatively limited number of atoms, the TS can then be an object of significant dimensionality embedded in the higher dimensional space in which the PES is defined. Additional complexity can arise from the PES being very rugged (e.g., the many local metastable configurations of a liquid will correspond to local minima on the PES representing the liquid). We shall also see in the following that a "good" choice of TS might need to take entropy into account (thus really defining a free energy surface instead of the original PES). These complexities are often less relevant to the most important (the slow) catalytic processes, which typically happen on relatively smooth PESs and have high energy barriers compared with many other types of atomic-scale processes. We shall therefore not go into any detail with such additional complexities.

The division of the PES into regions in Figure 4.1 is quite intuitive, and if the configuration of the atomistic system were entirely defined by the position of the atoms in space, this would be a perfectly useful definition of the reactant region, the TS, and the product region. It turns out, however, that the structure of a system does not entirely define its "physical state." We can think of this in terms of the question: What information do we need about the state of a system now in order to predict its configuration at a later point of time? If we think of this question in terms of Newton's 2nd law, which relates the rate of change in momenta, p, to the gradient of the potential

$$\frac{dp}{dt} = -\nabla V(x) \tag{4.2}$$

then we realize that at a given point of time, we need not only all positions of all the atoms, x, but also all of their momenta, p, (or equivalently their velocities), in order to calculate the configuration at a later time. For a full specification of the physical state of the system, we therefore need to specify the configuration in a $6N_{atoms}$-dimensional space made up of combined positions and momenta. We can thus think of Figure 4.1 as defining a division of configurations of combined positions and momenta into a reactant region and product region, the TS separating the two. The derivation of the TST rate constant in the following shall take advantage of an assumption of a thermally equilibrated reactant to separate the configuration space and the momentum space. The first assumption of TST is therefore that the energy of the reactant state is Boltzmann distributed. This is typically satisfied if the system has had enough time to thermally equilibrate in the reactant region. If the reactant region is unbounded, as could be the case with a gas-phase reactant, we assume that the reactant is impingent on the TS as a thermally equilibrated gas. This assumption would be fulfilled by a gas in thermal equilibrium, but is certainly not fulfilled by a molecular beam impingent on a surface, such as that discussed in Figure 2.3. Care therefore has to be taken when comparing such experiments directly to rate data or simulations.

The PES defines a unique correspondence between a potential and the nuclear positions of the system. When developing a rate theory based on the existence of a PES, we are therefore implicitly invoking the Born–Oppenheimer approximation. This approximation assumes that motion of the electrons is instantaneous compared with the motion of the nuclei, such that for whatever motion, we shall never move on an electronically excited state not corresponding to the ground-state potential energy. This is often a reasonable assumption, because the mass of any nucleus is on the order of 2×10^3–5×10^5 heavier than an electron. Since the forces exchanged between the nuclei and electrons are of similar size, the electrons can be assumed to be in their ground state in the electrostatic potential set up by the environment (typically the nuclei).

We shall in addition assume that the rate of quantum tunneling through the potential barriers is negligible compared with the rate obtained from the classical treatment. This is an assumption that generally always breaks down when the temperature becomes low enough. However, the typical crossover temperature below which the rate becomes dominated by quantum tunneling is very low (dependent on the barrier thickness and the masses of the tunneling particles).

Besides the three assumptions made earlier, one more central assumption is made in TST. The last assumption is that once the system attains a configuration in the TS with a velocity toward the product region, it will necessarily "react" and become the product. This sounds perhaps like an assumption that would necessarily always be fulfilled, but it is not. For such a configuration to lead to a reactive event, the system should not reenter the reactant region again shortly after moving into the product region. For some processes, many such "recrossings" occur, even with the best possible choice of TS. It is thus a significant approximation that one has to be aware of, since it leads to overestimations of the rate. This assumption, however, also leads to some very nice (variational) properties of TST, which can effectively be utilized for determining an appropriate separatrix as TS and to include corrections to beyond TST, as we shall discuss later in the chapter.

Finally, we define the "rate constant" for an elementary process as the rate of the process under the assumption that the system starts out by being in the reactant region. The rate constant is thus equal to the reaction rate, if the system is in the reactant configuration. This concept will be used to derive microkinetics in Chapter 5.

We are now ready to formulate TST. TST is the theory that, under the assumptions made earlier, decomposes the rate constant, k_{TST}, for an elementary process into the product of the probability of being in the TS, P_{TS}, and the rate, r_c, of passing through the TS in the direction toward the product region:

$$k_{TST} = P_{TS}r_c \qquad (4.3)$$

The rate for traversing the TS is of course infinite if the TS is infinitely thin (of one dimension less than the full configuration space) and the system is in the TS with a velocity of the system not parallel to the TS. Likewise, the probability for the system to be found in the TS relative to being found in the reactant region will be zero if the TS is infinitely thin. We therefore assume that the TS region has a small finite thickness, δx, and we determine the rate constant in the limit that the thickness of the TS goes to zero (Fig. 4.2). Now, the energy as a function of positions, x, and velocities, v, is (classically)

$$E(x,v) = E_{pot} + E_{kin}$$

$$E(x,v) = V(x) + \sum_i \frac{1}{2} m_i v_i^2 \qquad (4.4)$$

Using the assumption that the system is thermally equilibrated and therefore has Boltzmann-distributed energies, the probability density for finding the system at the point (x,v) in combined configuration and momentum space will be

$$P(x,v) \propto e^{-E(x,v)/k_B T} \qquad (4.5)$$

$$P(x,v) \propto e^{-V(x)/k_B T} \cdot e^{-\sum_i \frac{1}{2} m_i v_i^2 / k_B T} \qquad (4.6)$$

$$P(x,v) \propto P_x(x) \cdot P_v(v) \qquad (4.7)$$

Since the potential and kinetic energies are additive (the potential is velocity independent), the probability density for finding the system at a point (x,v) becomes

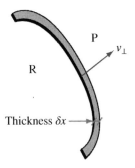

FIGURE 4.2 Definition of the vicinal region of the transition state and the perpendicular forward velocity. Here "R" designates the reactant region and "P" the product region.

separable into a product of two probability densities, one of which only depends on position and the other only depending on velocity. The kinetic energy is therefore not only Boltzmann distributed for the total system (meaning *averaged* over an entire PES) but also Boltzmann distributed in every single point, x, on the PES.

The rate at which the system traverses the infinitesimal region of thickness, δx, around the TS is given by the average velocity orthogonal to the configuration space TS (see Fig. 4.2):

$$r_{c,\delta x} = \frac{|v_\perp|}{\delta x} \tag{4.8}$$

The average orthogonal velocity can be evaluated directly from the Boltzmann distribution:

$$|v_\perp| = |v \cdot n(x)| \tag{4.9}$$

$$|v_\perp| = \frac{\int_{TS}\int_0^\infty |v \cdot n(x)| \cdot e^{-V(x)/k_B T} \cdot e^{-\sum_i \frac{1}{2}m_i v_i^2 / k_B T} \, dv \, dx}{\int_{TS} e^{-V(x)/k_B T} \, dx \cdot \int_{-\infty}^\infty e^{-\sum_i \frac{1}{2}m_i v_i^2 / k_B T} \, dv} \tag{4.10}$$

$$|v_\perp| = \sqrt{\frac{k_B T}{2\pi\mu}} \tag{4.11}$$

The variable, μ, introduced at this point is an effective mass of the system for the motion perpendicular to the configuration space TS.

The classical result of TST is then the definition of the TST rate constant, which is obtained by combining the Equations (4.3), (4.8), and (4.11):

$$k_{TST,\delta x}^{classical} = r_{c,\delta x} \cdot P_{TS\pm\frac{\delta x}{2}} = \frac{\sqrt{k_B T / 2\pi\mu}}{\delta x} \cdot P_{TS\pm\frac{\delta x}{2}} \tag{4.12}$$

We could now attempt to "classically" determine the probability of finding the system in the infinitesimal vicinity of thickness, $\pm\delta x/2$, around the TS surface as an expression like

$$P_{TS\pm\frac{\delta x}{2}} \neq \frac{\int_{\left(TS\pm\frac{\delta x}{2}\right)} e^{-V(x)/k_B T} \, dx}{\int_R e^{-V(x)/k_B T} \, dx} \tag{4.13}$$

where the denominator normalizes the probability, such that $P_{TS\pm\frac{\delta x}{2}}$ is the absolute probability for finding the system in the $\pm\delta x/2$ vicinity of the TS region. This, however, will not give the correct (quantum statistical mechanics) probability.

Equation (4.12) is therefore as far as we can get in the description of TST from a purely classical point of view. To derive the TST rate constant on a form that is useful for determining chemical rates, we now simply assume that the classically obtained lessons earlier also hold in a quantum setting. Specifically, we shall assume that with the same assumptions underlying the theory as the aforementioned, we can still talk about a "TS," which from a quantum mechanical viewpoint by no means is obvious. Secondly, we shall assume that we can still speak of a "rate of traversing the configuration space TS" and that it is given by the same expression as we derived classically. This assumption is not clearly evident either, even if a TS can be defined, since an equilibrium quantum system at nonzero temperature actually cannot really be said to have "a velocity," but should rather be described in terms of a thermally averaged superposition of stationary quantum states. The argument for this loose derivation is that it saves us considerable complexity as compared to an attempt on a fully quantum statistical mechanical exposition, and it actually gives the correct result. We shall thus proceed by assuming that the only necessary modification of the aforementioned classical derivation is to insert the correct probability of finding the system in the vicinity of the configuration space TS $P_{TS \pm \frac{\delta x}{2}}$. In quantum statistical mechanics, the equivalent of the classical probability density of finding the system in a given point is the probability of finding the system in a given state or set of states, and this probability is described by relative partition functions. We shall thus assume that we can write the probability of being in the δx vicinity of the configuration space TS compared to the reactant region, as a fraction of the partition functions:

$$P_{TS, \pm \frac{\delta x}{2}} = \frac{q^{TS \pm \frac{\delta x}{2}}}{q^{R}} = \frac{\sum_{j}^{(\text{all states } j \text{ in } TS \pm \frac{\delta x}{2})} e^{-\frac{\varepsilon_j}{k_B T}}}{\sum_{k}^{(\text{all states } k \text{ in } R)} e^{-\frac{\varepsilon_k}{k_B T}}} \tag{4.14}$$

That one can just ascribe energy levels to belong to a certain configuration space region is certainly not obvious, but we shall not argue further for this abstraction, except mentioning that the total number of states for a real system is incredibly large.

We need to take the limit as $\delta x \to 0$, and we shall therefore integrate out the degree of freedom in $q^{TS \pm \frac{\delta}{2}}$ perpendicular to the TS. This is the motion along the reaction path, which is an unbound mode. If δx is taken to be sufficiently small, the wave function for such a free motion (in a constant potential) can be described as a plane wave. The number of such quantum states in this degree of freedom can then be determined as box-quantized plane waves (particle-in-a-box solutions), which have the quantized energy spectrum

$$E_n = \frac{h^2 n^2}{8 \mu \delta x^2} \tag{4.15}$$

where μ is the effective mass for motion perpendicular to the TS. Now, the δx-partition function can be evaluated explicitly, giving

$$q^{\delta x} = \sum_{n=1}^{\infty} e^{-\frac{\varepsilon_n}{k_B T}} = \sum_{n=1}^{\infty} e^{-\frac{h^2 n^2}{8 \mu \delta x^2}/k_B T} \qquad (4.16)$$

If we assume the energy levels lie close compared to $k_B T$, we can rewrite the sum as an integral and evaluate this:

$$q^{\delta x} \approx \int_{n=0}^{\infty} e^{-\frac{h^2 n^2}{8 \mu \delta x^2}/k_B T} dn = \sqrt{\frac{2 \pi \mu k_B T}{h^2}} \cdot \delta x \qquad (4.17)$$

We can therefore write the probability of finding the system in the $\pm \delta/2$ vicinity of the TS as

$$P_{TS \pm \frac{\delta}{2}} = \frac{q^{TS \pm \frac{\delta}{2}}}{q^R} = \frac{q^{\delta x} \cdot q^{TS}}{q^R} = \sqrt{\frac{2 \pi \mu k_B T}{h^2}} \cdot \delta x \cdot \frac{q^{TS}}{q^R} \qquad (4.18)$$

This gives us the TST rate constant as the product of the rate at which the TS is traversed multiplied by the probability of the system to be found in the TS:

$$k_{TST} = r_{c,\delta} \cdot P_{TS \pm \frac{\delta}{2}} = \frac{\sqrt{k_B T / 2 \pi \mu}}{\delta x} \cdot \sqrt{\frac{2 \pi \mu k_B T}{h^2}} \cdot \delta x \cdot \frac{q^{TS}}{q^R}, \qquad (4.19)$$

which reduces to

$$k_{TST} = \frac{k_B T}{h} \cdot \frac{q^{TS}}{q^R} \qquad (4.20)$$

The partition function, q^{TS}, is now the partition function for being "in" the TS and not just in the $\pm \delta/2$ vicinity of the TS, thus now excluding the mode perpendicular to the TS.

The relative partition function expression, q^{TS}/q^R, is as mentioned earlier an expression for the relative probability of finding the system in the TS compared to finding it in the reactant state. Since one of the fundamental assumptions underlying the TST is that the distribution of states in the reactant region and the TS is thermally equilibrated, it is reasonable to set the relative partition functions equal to an equilibrium constant, which is defined through the differences in standard Gibbs energy between the TS and the reactant state:

$$K_{TST/R} = \frac{q^{TS}}{q^R} = e^{-\Delta G_{TS}°/k_B T} = e^{-(G_{TS}° - G_R°)/k_B T} \qquad (4.21)$$

The TST result is therefore often written on the form

$$k_{TST} = \frac{k_B T}{h} e^{-\Delta G_{TS}°/k_B T} \qquad (4.22)$$

or equivalently

$$k_{TST} = \frac{k_B T}{h} \cdot e^{\Delta S_{TS}°/k_B} \cdot e^{-\Delta H_{TS}°/k_B T} \qquad (4.23)$$

where $\Delta G°$, $\Delta S°$, and $\Delta H°$ are differences in standard Gibbs energy, entropy, and enthalpy between the TS and the reactant state, respectively. One should be careful when evaluating the standard entropy contribution in the TS to explicitly avoid including the entropy contribution from the translational motion along the reaction path (or more generally perpendicularly to the TS). The expressions in Equations (4.22) and (4.23) are often convenient for applications, since they allow for the straightforward inclusion of tabulated entropies and the extension of TST to open systems (e.g., gaseous reactants). Often, the change in entropy between the reactant and TS is relatively limited. This is particularly true for systems where there are no changes in unbound translational degrees of freedom, since free translations lead to by far the largest entropy contributions (as, e.g., the large entropy of gases). At catalytically relevant temperatures, $k_B T/h$ is of the order of 10^{13} s^{-1}. In the absence of large entropic effects, the prefactor should therefore be approximately 10^{13} s^{-1}. That experimental studies sometimes show significant variations in prefactors from 10^{13} s^{-1} can then be due to a number of different issues. Primarily, either the entropic contributions actually vary substantially between the reactant and the TS, or the differences from the 10^{13} s^{-1} "rule-of-thumb" result occur because the observation of something that is not an elementary rate is fitted to the Arrhenius expression corresponding to an elementary step. When a total rate is measured for a process that is actually a combination of several competing elementary steps, significant variations are observed in the experimentally observed activation barriers and prefactors as compared with the same quantities for the corresponding elementary steps.

In order to get a rough idea of what Equation (4.22) means, we will now show how the rate varies as a function of temperature for different values of the Gibbs free energy of activation, ΔG (see Fig. 4.3). As typical values, we choose 0.75, 1.5, and 2.25 eV, representing small, medium, and large barriers for heterogeneously catalyzed reactions, respectively. As we will show in Chapter 5, as a rule of thumb, a reasonable catalyst should have a reaction rate on the order of 1 site^{-1} s^{-1}. Although some catalysts also operate at rates as low as 10^{-2} site^{-1} s^{-1}, some others show high rates on the order of 100 site^{-1} s^{-1}. The catalytic activity thus depends both on the rate at which a reactant is turned into a product at an active site and the number of active sites available on the catalyst. We shall therefore focus here on the regime giving rates around 1 s^{-1}. As can be seen in Figure 4.3, barriers of 0.75 eV lead to rates of about 1 s^{-1} at room temperature. Moderate free energy barriers of 1.5 eV require temperatures of about 600 K in order to reach 1 s^{-1}, while large barriers require high reaction temperatures significantly over 800 K. The increase in rate with temperature is very steep in the beginning but flattens out at higher temperature. If we take 500 K, which is the region of many catalytic processes, one can deduce from Figure 4.3 that we can increase the rate by approximately 7.5 orders of magnitude when lowering the barrier from 1.5 to 0.75 eV. A change of 0.1 eV in barrier hence roughly leads to an order of magnitude change in rate.

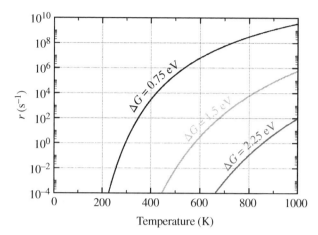

FIGURE 4.3 Rate plotted as a function of temperature for a Gibbs free energy of activation of $\Delta G = 0.75$ eV (black curve), $\Delta G = 1.5$ eV (light gray curve), and $\Delta G = 2.25$ eV (dark gray curve) as calculated using Equation (4.22).

Figure 4.4 shows a Gibbs free energy of activation, ΔG, plotted as a function of temperature for $r = 0.01$, $r = 1$, and $r = 100$ s^{-1}. If we take a rate of 0.01 s^{-1} as the limit (normally 1 s^{-1} is desirable), one can see the maximum allowed free energy barrier for a given temperature. For example, roughly 500 K is needed to yield a rate of 0.01 s^{-1} when the barrier is about 1.5 eV. It can also be seen that the rate increases substantially with increasing temperature, the increase being more pronounced at higher temperatures. The figure also explains why most heterogeneously catalyzed reactions proceed at temperatures between 400 and 600 K as these temperatures are needed in order to get reasonable reaction rates when barriers are somewhere between 1 and 1.5 eV.

Figure 4.5 shows the Gibbs free energy of activation, ΔG, as a function of rate for three different temperatures. Here, we can read of the rate that a certain barrier will give us for a certain temperature. At $\Delta G = 0$, r is only dependent on the prefactor that is roughly 10^{13} s^{-1}. Processes having very small barriers below 0.5 eV, which is often seen in surface diffusion (see also Chapter 2), are exceedingly fast even at room temperature and are hence usually equilibrated. We can also see that by increasing the temperature from 300 to 600 K (for $\Delta G = 0.75$ eV), we get an increase in rate from 1 to 10^7 s^{-1}. Likewise, this increase in temperature would allow the barrier to increase from 0.75 to 1.5 eV while preserving a rate of 1 s^{-1}. Figures 4.3, 4.4 and 4.5 thus allow us to obtain some rules of thumb on how the reaction rate is connected to the barrier and temperature.

For the study of trends in catalysis, one often looks at a given reaction and reaction pathway and varies the catalytic surface. Here, the variations in entropy from system to system are usually very small and give rise to variations in the rate by less than an order of magnitude. This should be compared to the changes in rates induced by varying the energy barriers of different reactions. A change in barrier of, for

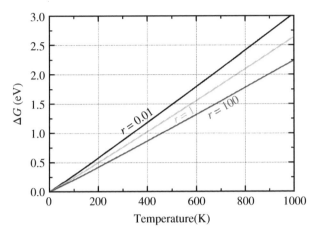

FIGURE 4.4   The Gibbs free energy of activation, ΔG, plotted as a function of temperature for $r = 0.01$ s^{-1} (black curve), $r = 1$ s^{-1} (light gray curve), and $r = 100$ s^{-1} (dark gray curve) as calculated using Equation (4.22).

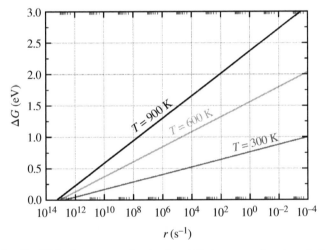

FIGURE 4.5   The Gibbs free energy of activation, ΔG, shown as a function of r for $T = 900$ K (black curve), $T = 600$ K (light gray curve), and $T = 300$ K (dark gray curve) as calculated using Equation (4.22).

example, 1 eV will at room temperature, for example, lead to a variation in the Boltzmann factor of approximately 10^{17}. Since the variation in activation barriers between neighboring metals in the periodic table is on the order of perhaps half an eV, it is often sufficient for the study of trends in heterogeneous catalysis to disregard effects of the varying prefactor and concentrate entirely on the variations in the reaction energetics.

4.3 RECROSSINGS AND VARIATIONAL TRANSITION STATE THEORY

We have not yet addressed how to determine a reasonable TS from knowledge of the PES. The so-called variational transition state theory (VTST) approach is the key tool for providing insight into how to make an optimal choice of TS.

It is assumed in TST that configurations that are found in the TS and have a velocity toward the product region will eventually end up in the product region. This means that cases where the supposed product crosses back into the reactant region are miscounted as "false positives" (see Fig. 4.6).

Since some trajectories that are counted as reactive events end up turning back to the reactant region, these lead to an overestimation of the TST rate constant as "false-positive" counts. Other reaction trajectories will cross the TS once, turn back to the reactant region, and finally cross the TS region again to end up in the product state, thus actually being reactive events. These trajectories also lead to an overestimation of the rate, since the single reactive event they represent leads to a double counting of the forward crossings of the TS. In fact it is observed that no matter how many times a trajectory crosses the TS before it turns into either products or reactants, the TST rate constant will be overestimated, since it assumes that every crossing of the TS toward the product region is one reactive event.

The overestimation of the TST rate constant associated with these recrossings may at first sound like a terrible nuisance, leading to a significant inaccuracy of the theory. However, the concept that the TST rate constant is always overestimated leads to a useful method for determining a "good TS," as it establishes a "variational principle" for the optimal dividing surface. A *variational principle* is a principle stemming from knowledge of some quantity always being strictly overestimated (or strictly underestimated). A *variational method* is then the derived method that utilizes the fact that since some quantity is always strictly overestimated (or strictly under-estimated), one can try many different variations of suggested optimal solutions, and

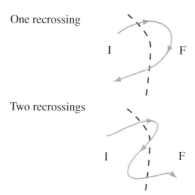

FIGURE 4.6 Two different kinds of recrossings that both lead to an overestimation of the rate constant.

the suggestion leading to the smallest value of the overestimated quantity (or largest value of the underestimated quantity) is then optimal in the tested space of trial solutions.

In general, we can here write

$$k_{\text{TST}} \geq k_{\text{exact}} \tag{4.24}$$

with equality holding only in the special case that a "trial" TS is a dividing surface with no recrossings at all. The variational method now stipulates that we should choose the TS such that k_{TST} is as small as possible. Such a TS is by definition the "optimal" TS to be used for a TST evaluation of the rate. By picking this optimal TS, we would obviously be overestimating the exact rate constant as little as possible. Unfortunately, however, there are very considerable difficulties in carrying out all possible variations of the dividing surface to find the absolutely optimal one. As we saw earlier in the chapter, the TS in configuration space could be a $(3N_{\text{atoms}} - 1)$-dimensional object, and even for relatively simple reactions with a few atoms, the TS can be almost impossible to even just represent—let alone carrying out—all possible variations. Extending rate theories for rare events to systems of many atoms is thus an active research area, and we point the reader to the Further Reading section at the end of the chapter. For many heterogeneous catalysis applications, one can find very reasonable TS, with a limited number of recrossings. This probably stems from the fact that the large barriers relevant for the slowest catalytic rates also often lead to significant trapping of the reactant as it traverses the TS.

A significant and straightforward improvement can be obtained over TST by including so-called dynamical corrections to the TST rate constant. This is practically done by first choosing a TS (perhaps variationally) and then evaluating by molecular dynamics the trajectories of a Boltzmann ensemble of starting points in the TS surface. These trajectories are followed for a very short time period compared to the time constant (inverse rate) of the reactive event, however, long enough to give a reasonable estimate of how many times they cross the TS before being thermally equilibrated in either of the reactant or product regions. The inclusion of such dynamical corrections to TST can for many processes essentially give the exact classical rate, as long as the assumptions mentioned in the beginning of the chapter are satisfied. This leads us to introducing the *transmission coefficient*, $\kappa(T)$, which is always smaller than one and which accounts for the recrossing corrections and therefore expresses the exact classical rate:

$$k_{\text{exact}} = \kappa(T) \cdot k_{\text{TST}} \tag{4.25}$$

$$k_{\text{exact}} = \kappa(T) \cdot \frac{k_B T}{h} \cdot e^{-\Delta G_{\text{TS}}^{\circ}/k_B T} \tag{4.26}$$

$$\kappa(T) < 1 \tag{4.27}$$

If one, for example, measures a TS entropy by fitting an expression such as

$$k = \frac{k_B T}{h} \cdot e^{-\Delta G_{TS}{}^\circ / k_B T} \tag{4.28}$$

to a measured rate constant, k, one should always remember that in fact the unknown transmission factor has been introduced into the experimentally measured entropy:

$$\kappa(T) \cdot \frac{k_B T}{h} \cdot e^{-\Delta G_{TS}{}^\circ / k_B T} = \frac{k_B T}{h} \cdot e^{(\Delta S_{TS}{}^\circ + k_B \ln \kappa(T))/k_B} \cdot e^{-\Delta H_{TS}{}^\circ / k_B T} \tag{4.29}$$

One can take another point of view and say that since the term $k_B \ln \kappa(T)$ for all intents and purposes behaves like a (perhaps slightly temperature-dependent) reduction of the standard entropy (since $\ln \kappa(T) < 0$), then we should consider it a genuine reduction of the entropy in the TS. We shall utilize this convention throughout the book, such that we consider Equation (4.28) an exact relation for the exact rate constant, k, and the exact standard Gibbs energy barrier, $\Delta G_{TS}{}^\circ$, which then again entails that we redefine the TS entropy as

$$S_{TS}{}^\circ = S'_{TS}{}^\circ + k_B \ln \kappa(T) \tag{4.30}$$

where $S'_{TS}{}^\circ$ is the classical (Gibbs grand canonical ensemble) entropy in the TS that is proportional to the logarithm of the number of accessible states and the term $k_B \ln \kappa(T)$, is a "recrossing correction" to the standard TS entropy.

Since the dynamical corrections involve molecular dynamics, or at least some sort of thermal sampling of a reasonably sized ensemble, these corrections can easily become rather computationally demanding to carry out. Often, it has been found fruitful to go in the opposite direction instead and use TST as a conceptual basis for making simpler estimates of the rate constants and base these on approximations to the true TS. One such approach, which has become extremely popular, is the HTST.

4.4 HARMONIC TRANSITION STATE THEORY

In HTST, a harmonic expansion of the PES is invoked both in the IS and in the saddle point separating the IS and the FS. The HTST is therefore applicable under the same general assumptions as mentioned for TST but further demands that the PES is smooth enough for a local harmonic expansion of the PES to be reasonable. This means that it is necessary that the potential is reasonably well represented by its second-order Taylor expansion around these two expansion configurations. The general idea is that the partition functions in Equation (4.20) can be evaluated analytically for the harmonic expansion of the PES around the expansion points. This leads to very simple expressions for the rate constants and gives reasonable rate constants for those systems for which the underlying assumptions are not violated. Adsorbates reacting with each other on surfaces have in a number of cases been shown to have rate constants that are surprisingly well represented by this relatively crude approximation to full TST.

The procedure for determining the HTST rate constant thus follows a series of well-defined steps. These are steps allowing the determination of a rate constant based on an extremely limited sampling of the PES. First, an IS is determined as the lowest energy point in the reactant region, for example, by direct structural optimization. Then, a dominant first-order saddle point on the PES needs to be determined. A number of practical methods exist for performing such saddle point searches. For the extended systems (surfaces) relevant for heterogeneous catalysis, one is usually limited to saddle point search methods, which only employ forces (first derivatives of the potential) and energies. A majority of studies in theoretical catalysis focus on predefined elementary reactions, for which the IS and FS are assumed to be known, and the objective of the search method is to find the saddle point between these states. For problems of this type, the nudged elastic band (NEB) method or one of the many derived methods is typically used. The NEB algorithm establishes a string of "images" of the system that lie approximately optimally along the minimal energy path (MEP) from the IS to the FS of the reaction. To use the NEB algorithm, one needs to *a priori* determine the FS as well as the IS. The saddle point is determined as the maximum energy configuration along the MEP. The TS in HTST is then the uniquely defined dividing surface, which is the hyperplane (a plane without curvature or so to say with a constant normal vector everywhere) going through the saddle point and which is perpendicular to the reaction path at the saddle point.

The next step is to perform a normal mode analysis, which is a method for finding the uncoupled orthogonal vibrational modes of the system. Expressed in coordinates, $q_{i,IS}$, from the IS along these D orthogonal modes, a harmonic expansion of the potential in the reactant region can then be established as

$$V_R\left(q_{IS}\right) = V_{IS} + \sum_{i=1}^{D} \frac{1}{2} k_{i,\,IS} q_{i,\,IS}^2 \tag{4.31}$$

The potential in the TS hyperplane is similarly

$$V_{TS}\left(q_{SP}\right) = V_{SP} + \sum_{i=1}^{D-1} \frac{1}{2} k_{i,\,SP} q_{i,\,SP}^2 \tag{4.32}$$

The force constants, k_i, are also obtained from the normal mode analysis and used to obtain the vibrational frequencies, v_i, for the different modes with $v_i = \dfrac{1}{2\pi} \cdot \sqrt{k_i/\mu_i}$ for the vibrational eigenmode i corresponding to an effective mass in the vibrational direction of μ_i. The partition functions for the potential expansions in Equations (4.31) and (4.32) can then be evaluated analytically and introduced in the expression for the rate constant in Equation (4.20). In the harmonic approximation, we can then write up the probability of finding the system in the TS, q^{TS}/q^R, by summing over all the vibrational levels and all vibrational modes. Since the energy levels of the harmonic oscillator mode with frequency v_i are given by

$$\varepsilon_n = hv_i\left(n + \frac{1}{2}\right), \quad n = 0, 1, 2, \ldots \tag{4.33}$$

we get a partition function for the harmonic mode i to be

$$q_i = \sum_{n=0}^{\infty} e^{-\frac{\varepsilon_n}{k_B T}} = \sum_{n=0}^{\infty} e^{-\frac{hv_i\left(n+\frac{1}{2}\right)}{k_B T}} = e^{-\frac{hv_i/2}{k_B T}} \sum_{n=0}^{\infty} e^{-\frac{hv_i}{k_B T}n} \tag{4.34}$$

We utilize that geometric series can be reduced to closed expressions by observing that

$$\sum_{n=0}^{\infty} z^n = 1 + z + z^2 + z^3 + z^4 + \cdots \tag{4.35}$$

and

$$z\sum_{n=0}^{\infty} z^n = z + z^2 + z^3 + z^4 + \cdots \tag{4.36}$$

such that when we subtract these two series, we get the result of "1":

$$\sum_{n=0}^{\infty} z^n - z\sum_{n=0}^{\infty} z^n = 1 + z + z^2 + z^3 + z^4 + \cdots - \left(z + z^2 + z^3 + z^4 + \cdots\right) = 1 \tag{4.37}$$

We therefore have

$$(1 - z)\sum_{n=0}^{\infty} z^n = 1, \tag{4.38}$$

which implies that

$$\sum_{n=0}^{\infty} z^n = (1 - z)^{-1} \tag{4.39}$$

Setting $z = e^{-\frac{hv_i}{k_B T}}$ in Equation (4.34), we then obtain

$$q_i = e^{-\frac{hv_i/2}{k_B T}} \sum_{n=0}^{\infty} e^{-\frac{hv_i}{k_B T}n} = e^{-\frac{hv_i/2}{k_B T}} \left(1 - e^{-\frac{hv_i}{k_B T}}\right)^{-1} \tag{4.40}$$

The D vibrational modes in the IS give a reactant region partition function on the form

$$q^{R}\Big|_{\text{Harmonic}} = \sum_{n_1,n_2,\ldots,n_D=0}^{\infty} e^{-\frac{V_{\text{IS}}+\varepsilon_{n_1}+\varepsilon_{n_2}+\cdots+\varepsilon_{n_D}}{k_{\text{B}}T}} \tag{4.41}$$

This separates into a product of individual vibration mode partition functions:

$$q^{R}\Big|_{\text{Harmonic}} = e^{-\frac{V_{\text{IS}}}{k_{\text{B}}T}} \cdot \sum_{n_1=0}^{\infty} e^{-\frac{\varepsilon_{n_1}}{k_{\text{B}}T}} \cdot \sum_{n_2=0}^{\infty} e^{-\frac{\varepsilon_{n_2}}{k_{\text{B}}T}} \cdots \sum_{n_D=0}^{\infty} e^{-\frac{\varepsilon_{n_D}}{k_{\text{B}}T}} \tag{4.42}$$

such that the total reactant region partition function becomes

$$q^{R}\Big|_{\text{Harmonic}} = e^{-\frac{V_{\text{IS}}}{k_{\text{B}}T}} \cdot \prod_{i=1}^{D} e^{-\frac{h\nu_i^{\text{IS}}/2}{k_{\text{B}}T}} \left(1 - e^{-\frac{h\nu_i^{\text{IS}}}{k_{\text{B}}T}}\right)^{-1} \tag{4.43}$$

We shall conveniently rewrite this as

$$q^{R}\Big|_{\text{Harmonic}} = e^{-\frac{V_{\text{IS}}+\sum_{i=1}^{D} h\nu_i^{\text{IS}}/2}{k_{\text{B}}T}} \cdot \prod_{i=1}^{D} \left(1 - e^{-\frac{h\nu_i^{\text{IS}}}{k_{\text{B}}T}}\right)^{-1} \tag{4.44}$$

Similarly, the TS partition function is given by the $(D-1)$ frequencies in the saddle point as

$$q^{\text{TS}}\Big|_{\text{Harmonic}} = e^{-\frac{V_{\text{SP}}+\sum_{j=1}^{D-1} h\nu_j^{\text{SP}}/2}{k_{\text{B}}T}} \cdot \prod_{j=1}^{D-1} \left(1 - e^{-\frac{h\nu_j^{\text{SP}}}{k_{\text{B}}T}}\right)^{-1} \tag{4.45}$$

We can now use Equation (4.20) with the harmonic partition functions

$$k_{\text{HTST}} = \frac{k_{\text{B}}T}{h} \cdot \frac{q^{\text{TS}}}{q^{R}}\Big|_{\text{Harmonic}} \tag{4.46}$$

to determine the HTST rate constant:

$$k_{\text{HTST}} = \frac{k_{\text{B}}T}{h} \cdot \frac{\prod_{i=1}^{D}\left(1 - e^{-\frac{h\nu_i^{\text{IS}}}{k_{\text{B}}T}}\right)}{\prod_{j=1}^{D-1}\left(1 - e^{-\frac{h\nu_j^{\text{SP}}}{k_{\text{B}}T}}\right)} \cdot e^{-\frac{\Delta E_a}{k_{\text{B}}T}} \tag{4.47}$$

where the product ratio represents the thermal contributions to the entropy (and the internal energy corrections to the activation energy), and we have defined the zero-point-energy-corrected activation energy as

$$\Delta E_a = \left(V_{SP} + \sum_{j=1}^{D-1} h\nu_j^{SP}/2\right) - \left(V_{IS} + \sum_{i=1}^{D} h\nu_i^{IS}/2\right) \tag{4.48}$$

If the frequencies are small, such that $h\nu_i \ll k_B T$ for all i, we can make the Taylor expansion

$$e^x = 1 + x + \frac{1}{2}x^2 + \frac{1}{3!}x^3 + \cdots \tag{4.49}$$

of the exponentials in the product ratio (to first order):

$$\left(1 - e^{-\frac{h\nu}{k_B T}}\right) \approx 1 - 1 + \frac{h\nu}{k_B T} = \frac{h}{k_B T}\nu \tag{4.50}$$

The HTST rate constant is then approximately (assuming that $h\nu_i \ll k_B T$ for all i)

$$k_{HTST} \approx \frac{\prod_{i=1}^{D} \nu_i^{IS}}{\prod_{j=1}^{D-1} \nu_j^{SP}} \cdot e^{-\Delta E_a/k_B T} \tag{4.51}$$

The HTST rate constant in this case therefore reduces to an Arrhenius expression:

$$k_{Arrhenius} = \upsilon \cdot e^{-\Delta E_a/k_B T} \tag{4.52}$$

Since in Equation (4.51) there is one frequency more in the numerator than in the denominator, the prefactor, υ, is often practically thought of as a frequency of an atomic vibration and is therefore also referred to as an "attempt frequency." Typically, however, for slow surface reactions of relevance in heterogeneous catalysis, not all frequencies will be small, and the assumption $h\nu_i \ll k_B T$ is therefore very poorly satisfied. The simplified expression in Equation (4.51) will then lead to unacceptably large errors. We shall therefore generally discourage the use of Equation (4.51) and rather rely on the "exact" HTST rate constant expression in Equation (4.47), remembering that the activation energy should be zero-point energy corrected, which in the presence of large frequencies can lead to significant corrections (see Chapter 3).

Many adsorbate and bulk systems fulfill the assumptions for HTST to apply. For strongly bonded systems, the PES is often sufficiently harmonic in the IS and in the saddle point. This is perhaps due to the rather large energy barriers involved in most surface processes. Almost always, the most important processes are very slow compared to molecular vibrations. This can be seen as natural, since strong bonds lead to attempt frequencies around 10^{13} s^{-1}. If the energy barrier is not substantial, the rate automatically becomes extremely high, and the reactant and product would then turn out to establish thermal equilibrium.

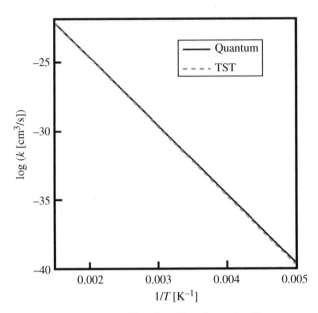

FIGURE 4.7   Difference between HTST and multiconfiguration Hartree approach. Adapted from van Harrevelt et al. (2005).

For thin enough barriers and at low enough temperatures, the transmission coefficient will get quantum tunneling corrections and can actually grow to become (much) larger than one, even in presence of recrossings. This is due to quantum tunneling through the reaction barrier. The crossover temperature at which the reaction rate is no longer dominated by the classical rate expressions derived earlier, but by thermally activated quantum tunneling, is typically significantly lower than room temperature. Since lighter atoms tunnel more readily, hydrogen is one of the few atoms, which can tunnel at reasonably high temperatures (e.g., at room temperature), dependent on the barrier thickness. A fully satisfactory (variational) quantum TST has still not been devised. This is primarily because the reasonably complete description of quantum tunneling requires solution of the full Schrödinger equation.

Figure 4.7 shows results from an HTST calculation of the rate constant for N_2 dissociation over a ruthenium surface as compared to results obtained using the much more advanced and significantly more computationally demanding "multiconfiguration Hartree approach," which explicitly takes quantum tunneling effects into account. It is observed that the two lines representing HTST and the multiconfiguration Hartree approach are essentially indistinguishable down to low temperatures ($1/T = 0.005$ K^{-1} corresponds to $T = 200$ K). The result illustrates that for this reaction not only are the thermal quantum tunneling corrections to the TST rate constant small but it is also that the harmonic approximation to TST performs extraordinarily well for this process.

Some important systems, which certainly do not fulfill the assumptions of HTST, are liquid-phase reactions. For these, many different reaction paths are typically present, and the PESs in vicinity of the first-order saddle points are not adequately well represented by a harmonic expansion (the first-order saddle points are not well separated from the second-order saddle points on the scale of $k_B T$). This problem is difficult to correct for. Other important systems, to which HTST does not directly apply in the form derived earlier, are those where the reactant is a gas-phase molecule. In the gas phase, there are zero modes such as translation and rotation, and these lead to totally different partition functions than those obtained from a normal mode analysis. One extends HTST to this type of reaction simply by inserting the gas-phase partition function into Equation (4.20) instead of the harmonic expansion or by inserting tabulated entropies associated with translation and rotation instead of the vibrations into Equation (4.23). One can thereby in a simple manner still use the results of HTST to obtain the relevant rate constant.

REFERENCE

van Harrevelt R, Honkala K, Nørskov JK, Manthe U. The reaction rate for dissociative adsorption of N_2 on stepped Ru(0001): six-dimensional quantum calculations. J Chem Phys 2005; 123:234702.

FURTHER READING

Bennett CH. Exact defect calculations in model substances. In: Nowick AS, Burton JJ, editors. *Algorithms for Chemical Computation*. ACS Symposium Series No. 46, p 63. Washington, DC: American Chemical Society; 1977.

Chandler D. Statistical-mechanics of isomerization dynamics in liquids and transition-state approximation. J Chem Phys 1978;68:2959.

Eyring H. The activated complex in chemical reactions. J Chem Phys 1935;3:107.

Eyring H. The theory of absolute reaction rates. Trans Faraday Soc 1938;34:41.

Henkelman G, Jónsson H. Improved tangent estimate in the nudged elastic band method for finding. J Chem Phys 2000;113:9978.

Henkelman G, Uberuaga BP, Jónsson H. A climbing image nudged elastic band method for finding minimal energy paths and saddle points. J Chem Phys 2000;113:9901.

Jóhannesson GH, Jónsson H. Optimization of hyperplanar transition states. J Chem Phys 2001;115:9644.

Jónsson H, Mills G, Jacobsen KW. Nudged elastic band method for finding minimum energy paths of transitions. In: Berne BJ, Ciccotti G, Coker DF, editors. *Classic and Quantum Dynamics in Condensed Phase Simulations*. Singapore: World Scientific; 1998.

Olsen RA, Kroes GJ, Henkelman G, Arnaldsson A, Jonsson H. Comparison of methods for finding saddle points without knowledge of the final states. J Chem Phys 2004;121:9776.

Perez D, Uberuaga BP, Shim Y, Amar JG, Voter AF. Accelerated molecular dynamics methods: Introduction and recent developments. Ann Rep Comp Chem 2009;5:79–98.

Wigner E. The transition state method. Trans Faraday Soc 1938;34:29.

5

KINETICS

In the previous chapters, we have dealt with the methodology for describing equilibria of chemical reactions as well as the methodology for obtaining rate constants for elementary reactions. In order to describe the total reaction rate of a set of coupled reactions (and a heterogeneously catalytic reaction is always a coupling of several elementary reaction steps), we shall now combine these descriptions to arrive at a logical treatment of coupled elementary steps. This description is referred to as "microkinetic modeling."

5.1 MICROKINETIC MODELING

We saw in Chapter 3 how it was relatively simple to obtain an expression for the configurational entropy for a given coverage of an adsorbate from a combinatorial argument for randomly distributed adsorbates. In reality, adsorbates of course do interact as we also witnessed in Chapter 3. Adsorbates on an ordered surface will during their motion on the surface attempt to minimize their free energy. If the thermal fluctuations are small enough compared with the corrugation of the potential energy surface, but large enough that the adsorbates can slowly diffuse around, then adsorbate structuring on the surface will occur. The adsorbates can then form periodic structures (to minimize their mutual repulsion), islands (to maximize their mutual attraction), undergo phase separations, or form various other types of short-range

Fundamental Concepts in Heterogeneous Catalysis, First Edition. Jens K. Nørskov,
Felix Studt, Frank Abild-Pedersen and Thomas Bligaard.
© 2014 John Wiley & Sons, Inc. Published 2014 by John Wiley & Sons, Inc.

or long-range ordering. This complexity is often difficult to treat exactly, and the solution of such problems with many interacting bodies is a key research area of statistical mechanics. In statistical mechanics, an often-utilized first approximation to a solution is the so-called mean field model.

In a mean field model, one replaces all the detailed interactions between any one body and the rest of the system with an average or "effective" interaction. This replacement turns a many-body problem into a set of one-body problems. The same approach is used to model catalytic reactions on surfaces. In catalysis, the term "mean field microkinetic modeling," however, commonly refers to the specific microkinetic model in which all repulsive (or attractive) interactions between adsorbates have been removed. This strategy has (perhaps surprisingly) in a large number of studies turned out to give remarkably good agreement with experimental studies. Significant care has to be taken, however, since adsorbate interactions can be quite significant, as we saw in Chapter 2.

5.2 MICROKINETICS OF ELEMENTARY SURFACE PROCESSES

Let's assume we have determined a rate constant, k_-, for the desorption of species A from a surface. Recall that the rate constant is the rate for the desorption process, assuming that the adsorbate (A) is in the reactant (i.e., adsorbed) state. The rate at which desorption occurs from a given site on the surface is then proportional to the product of the rate constant and the probability that a given site is occupied. The probability that a site is occupied is equal to the fractional coverage of the adsorbate on the surface, θ_A. So we arrive at the expression

$$r_{desorption} = k_- \cdot \theta_A \tag{5.1}$$

Likewise, we would expect $r_{adsorption}$ to be proportional to the probability that a site is unoccupied, θ_*. We saw in Chapter 3 that at equilibrium this process is described by the equation

$$\theta_A = K_{ads} p_A \cdot \theta_* \tag{5.2}$$

When adsorption and desorption are equilibrated, the rates for adsorption and desorption are equal in magnitude:

$$r_{adsorption} = r_{desorption} \tag{5.3}$$

So, at equilibrium, we must have

$$r_{adsorption} = k_- K_{ads} p_A \cdot \theta_* = k_+ p_A \cdot \theta_* \tag{5.4}$$

We can see that the following relation must hold:

$$K_{ads} = \frac{k_+}{k_-} \tag{5.5}$$

We can think of the Equation (5.4) as expressing the adsorption rate as the product of 2 terms: (i) the rate constant k_+ at a flux of A toward the surface corresponding to standard pressure (1 bar), p_A, which describes the flux variation due to variations in the pressure relative to the standard pressure, and (ii) the coverage of free sites, θ_*, which is the probability that the site we are trying to adsorb in is a free site. Since this point of view makes sense even when the system is not in equilibrium, we shall assume Equation (5.4) holds for nonequilibrium situations.

The expression (5.5), which states that the equilibrium constant for an elementary reaction (here adsorption/desorption) is equal to the forward divided by the backward rate constant, is actually very general. We observe that

$$K = e^{-\Delta G^\circ / k_B T} = \frac{k_+}{k_-} = \frac{\dfrac{k_B T}{h} e^{-(G_{+TS}{}^\circ - G_I{}^\circ)/k_B T}}{\dfrac{k_B T}{h} e^{-(G_{-TS}{}^\circ - G_F{}^\circ)/k_B T}} = e^{-(G_F{}^\circ - G_I{}^\circ)/k_B T},$$

where the last equality only holds if the forward and the backward $G_{TS}{}^\circ$ terms cancel each other. At equilibrium, they do, meaning that the forward and the backward reactions go through the exact same transition state. This is a manifestation of the *principle of microscopic reversibility*. This principle (originally due to Ludwig Boltzmann and derived for gas-phase collisions) is a result of the time-reversal symmetry of the underlying mechanical laws governing the dynamics of the system (whether we take those laws to be Newton's 2nd, the time-dependent Schrödinger equation, or the Dirac equation, they all manifest time-reversal symmetry). This leads to what is called *the principle of detailed balance*—the concept that for a set of multiple coupled elementary reactions at equilibrium, the forward rate of each elementary reaction will be identical to the backward rate of the same elementary reaction. Since Equation (5.5) thus is a manifestation of a deeper-lying principle, one way to view the "derivation" of Equation (5.5) is that we have chosen rate expressions for adsorption and desorption that conform to our earlier calculation of the configurational entropy of adsorption in Appendix 3.3.

The rate expressions for surface processes thus take the same form as well-known expressions for chemical reactions in gas phase or in solution. Whereas the activities of gas- or liquid-phase reactants are expressed as pressures or concentrations, the activities of surface reactants are expressed in terms of the fractional coverages of adsorbates and of free sites. For surface processes, we can define reaction fractions by taking the product of the activities of the left-hand side of a reaction equation and dividing it by the right-hand side. For example, with the adsorption–desorption reaction given by

$$A + * \rightleftarrows A * \tag{5.6}$$

we can associate the reaction fraction, $\theta_A / p_A \theta_*$, and at equilibrium, we have

$$\left. \frac{\theta_A}{p_A \theta_*} \right|_{Eq} = K_{ads} = e^{-\frac{\Delta G^\circ_{ads}}{k_B T}} \tag{5.7}$$

As we shall see, it is convenient to define the "approach to equilibrium" (also sometimes called the "reversibility") for an elementary step or for an overall reaction as the rate of the backward reaction divided by the rate of the forward reaction. We see that this can also be thought of as the reaction fraction divided by the equilibrium constant. For the adsorption reaction, the approach to equilibrium is

$$\gamma_{ads} = \frac{\theta_A}{p_A \theta_*} K_{ads}^{-1} \tag{5.8}$$

The approach to equilibrium is thus a positive quantity, which satisfies the following statements:

$$\gamma < 1: \text{ The net reaction is in the forward direction.} \tag{5.9}$$

$$\gamma = 1: \text{ The reaction is in equilibrium.} \tag{5.10}$$

$$\gamma > 1: \text{ The net reaction is in the backward direction.} \tag{5.11}$$

The point of view taken earlier suggests a straightforward extension to other elementary reactions of importance in describing catalytic surface processes. If we look at dissociative chemisorption (as an elementary reaction)

$$A_2 + 2* \rightarrow 2A* \tag{5.12}$$

we would expect

$$r_{diss} = k_{diss} \cdot p_{A_2} \cdot \theta_*^2 \tag{5.13}$$

And for the reverse process of associative desorption,

$$2A* \rightarrow A_2 + 2* \tag{5.14}$$

will have a rate expression given by

$$r_{ass} = k_{ass} \cdot \theta_A^2 \tag{5.15}$$

In these expressions, the squared coverage, θ_*^2 (or θ_A^2), can be thought of as describing the probability of finding two empty sites (or two sites both with an atom A) next to each other (assuming that the free sites or adsorbates are randomly distributed on the surface) as a prerequisite for the reaction to occur.

For a surface diffusion process of an adsorbate, A, from one site to a neighboring site, we would expect it could be described by an elementary process step such as

$$A*+* \rightarrow *+A* \tag{5.16}$$

and that it would have a rate expression given by

$$r_{\text{dif}} = k_{\text{dif}} \cdot \theta_A \cdot \theta_* \tag{5.17}$$

The coverages in this expression now represent the product of probabilities of finding an adsorbate A in the site we are looking at and an unoccupied site next to it. For a surface-mediated coupling/scission reaction

$$A* + B* \leftrightarrow AB* + * \tag{5.18}$$

we obtain the rate expressions

$$r_{\text{coupling}} = k_{\text{coupling}} \cdot \theta_A \cdot \theta_B \tag{5.19}$$

$$r_{\text{scission}} = k_{\text{scission}} \cdot \theta_{AB} \cdot \theta_* \tag{5.20}$$

For the very similar disproportionation reaction

$$AB* + C* \leftrightarrow A* + BC* \tag{5.21}$$

we obtain the forward rate expression

$$r_{\text{disprop}} = k_{\text{disprop}} \cdot \theta_{AB} \cdot \theta_C \tag{5.22}$$

For all the aforementioned two-site processes (dissociation, association, diffusion, coupling, scission, and disproportionation), one should remember that on a two-dimensional surface (or along a one-dimensional step), there is rarely only one neighboring site to react with. The probability of finding, for example, an empty neighboring site is therefore proportional to the coverage of empty sites. If an adsorbate in one site has, for example, six neighbors with whom it can react along similar reaction paths, then a factor of 6 should be included in the rate constant.

This can be thought of within the framework of transition state theory (TST) (Chapter 4) as the configuration integral in the transition state running over the whole surface separating the reactant region from the six distinct product regions. Whereas the factor corresponding to the number of neighbors is therefore intrinsically taken into account in the general formulation of TST, it is not automatically taken into account in harmonic transition state theory (HTST), in which a harmonic expansion of the transition state configuration integral is carried out in only one first-order saddle point on the potential energy surface. Using HTST in combination with microkinetics, one should therefore in principle always remember to include the number of equivalent paths, $N_{\text{equivalent paths}}$, as an extra factor in the appropriate rate constants. In terms of full TST, this corresponds to an increase in the entropy of the transition state by $k_B T \cdot \ln(N_{\text{equivalent paths}})$. Since this is on the order of a couple of $k_B T$s and therefore significantly smaller than typical errors in the energetics, the correction is often not taken into account in actual applications. As the quantitative analysis of catalytic reactions becomes increasingly more accurate and in closer correspondence with experiments, one may, however, eventually wish to include such corrections.

APPARENT ACTIVATION ENERGY OF AN ELEMENTARY REACTION STEP

An often employed method for analyzing experimentally obtained reaction rates is to derive the so-called apparent activation energy. One assumes an Arrhenius expression ($r = A \cdot e^{-E_A/k_B T}$) where the prefactor, A, and activation energy, E_A, are temperature independent. After a little algebra, one arrives at the expression

$$E_A^{\text{apparent}} = -\frac{\partial(\ln r)}{\partial(1/k_B T)} \tag{5.23}$$

This strategy typically gives reasonable results for reactions taking place in gas phase or in solution, where the pressures or the concentrations can be controlled at varying temperatures. On surfaces, however, one needs to be careful when using this procedure. Typically, it is still the pressures or the concentrations of the reactants that are controllable. The coverages, which play the role of activities for surface reactions, are very difficult to control as the temperature varies. If we therefore are trying to find the apparent activation energy of an elementary surface reaction (e.g., the disproportionation reaction)

$$E_A^{\text{apparent}} = -\frac{\partial\left(\ln r_{\text{disprop}}\right)}{\partial(1/k_B T)} = -\frac{\partial\left(\ln k_{\text{disprop}} + \ln\theta_{\text{AB}} + \ln\theta_C\right)}{\partial(1/k_B T)} \tag{5.24}$$

in which we would expect (approximately) $k_{\text{disprop}} = A \cdot e^{-E_A/k_B T}$, then we obtain

$$E_A^{\text{apparent}} = E_A - \frac{\partial\left(\ln\theta_{\text{AB}} + \ln\theta_C\right)}{\partial(1/k_B T)} \tag{5.25}$$

There can thus be a discrepancy between the calculated apparent activation barrier and the actual activation barrier for the process. If the coverages θ_{AB} and θ_C are close to one and therefore do not vary much with temperature, then the correction is small. If, however, the coverages of AB and C are *small* and *in equilibrium* with AB and C in the gas phase, the coverages are $\theta_{\text{AB}} \approx K_{\text{ads, AB}}$ and $\theta_C \approx K_{\text{ads, C}}$ (as given by the Langmuir isotherm; see Chapter 3), such that

$$E_A^{\text{apparent}} \approx E_A + E_{\text{ads, AB}} + E_{\text{ads, C}} \tag{5.26}$$

The reaction barrier, E_A, that is really measured from the adsorbed states of AB and C will thus apparently be the barrier measured from the gas phase when the coverages are *small* and *in equilibrium* with the gas-phase reactants.

5.3   THE MICROKINETICS OF SEVERAL COUPLED ELEMENTARY SURFACE PROCESSES

We shall now utilize the microkinetic expressions for individual elementary surface processes, as described in the previous section, in order to establish a model for describing full "catalytic cycles." A heterogeneous catalytic process always includes a number of elementary steps happening in sequence one after the other. If the reaction occurs over a surface, for example, the reactants first need to be adsorbed on the surface, then move into the vicinity of each other through diffusion processes, then react with each other forming the products, and finally desorb. Sometimes, the adsorbed reactants need to undergo various activation steps before being able to react with each other.

In order for the process to be catalytic, the catalyst should not be consumed in the process. A process comprising a number of elementary steps can therefore be written up in such a way that the reactants have been turned into products according to an integer times the reaction stoichiometry. The surface after having taken part in a number of reactions has been reinstated in its original configuration. Such a series of elementary reaction steps is referred to as a "catalytic cycle." When analyzing catalytic processes it is important always to look at the full catalytic cycle, as it is otherwise very easy to misinterpret key aspects of the process.

Let us first focus on the simple catalytic cycle, whereby two reactant gases A_2 and B react to form the gas AB. We assume that the process proceeds in the two elementary steps

$$A_2 + 2* \rightarrow 2A* \tag{5.27}$$

$$A* + B \rightarrow AB + * \tag{5.28}$$

and that the rate constants for the elementary process have already been determined. In the second step, which the cycle has to undergo twice for every time the first step takes place, the reactant B reacts with the adsorbed $A*$ straight from the gas phase. This is a so-called Eley–Rideal step, and it occurs very rarely in heterogeneous catalytic reactions between gas-phase reactants. In electrochemistry, it often occurs, however, with B having an electrical charge. In the electrochemistry field, this is referred to as a Heyrovský mechanism. We here employ it with the sole purpose of keeping the kinetics simple. We can now write up the reaction rate expressions

$$R_1 = r_1 - r_{-1} = k_1 \cdot p_{A_2} \cdot \theta_*^2 - k_{-1} \cdot \theta_A^2 \tag{5.29}$$

$$R_2 = r_2 - r_{-2} = k_2 \cdot p_B \cdot \theta_A - k_{-2} \cdot p_{AB} \cdot \theta_* \tag{5.30}$$

In these expressions, R_1 describes the net rate of A_2 removal, and R_2 describes the net rate of AB formation. Since both of the two elementary processes need to occur in order for the product to be formed and since step 1 changes the coverage that goes into step 2, the two rate expressions must be solved simultaneously in order to obtain the overall reaction rate. If we think of the net rates in terms of what surface species they create and remove, we see that reaction 1 (eq. 5.27) creates 2A*, while reaction 2 (eq. 5.28) removes 1A*. This establishes a differential equation for the time development of the coverage of A:

$$\frac{\partial \theta_A}{\partial t} = 2R_1 - R_2 = 2k_1 \cdot p_{A_2} \cdot \theta_*^2 - 2k_{-1} \cdot \theta_A^2 - k_2 \cdot p_B \cdot \theta_A + k_{-2} \cdot p_{AB} \cdot \theta_* \quad (5.31)$$

Though there are two coverages, that is, that of free sites and that of A, one differential equation is enough to completely specify the reaction, since the coverage of free sites follows from the coverage of A through the site conservation rule:

$$\sum_i \theta_i = 1 \tag{5.32}$$

Often, for example, we are not actually interested in the exact temporal behavior at all points of a reactor, but perhaps we aim to model and thereby compare how various catalysts behave under similar conditions. In order to do this, we employ the *steady reaction conditions assumption*, which assume that we are describing some point or slice of the catalytic reactor, where the pressure and the temperature are given and constant. We shall throughout the remainder of the book always make the steady reaction conditions assumption. By employing this assumption, the differential equation (5.31) is simplified by the fact that the rate constants and reactant and product pressures no longer are time dependent. The coverages, however, are of course still time dependent.

When there is only one independent variable (coverage) involved, such as in Equation (5.31), the general behavior is that the system moves toward a steady state. We see that in Equation (5.31), the rate of change of the coverage of A becomes negative if A gets close enough to 1 and positive if the coverage of A is close to zero. The coverage of A will thus move in the direction of lowering the rate of change of the coverage of A, and the coverage of A will asymptotically move toward the coverage at which the rate is zero.

SOLVING THE KINETIC MODEL

Equation (5.31) may appear to be a relatively simple equation, but it can in fact represent rather significant complexity. In an industrial chemical reactor, for example, there are pressure and temperature gradients, such that all reactant and product pressures, rate constants, and coverages vary through the reactor, and the previous differential equation therefore might need to be solved as a function of position in the reactor. Furthermore, the pressures and the temperature can vary in time and depend on the reaction rate, whereby all the pressures, rate constants, and coverages might have to be solved for as a function of time and in many positions through the reactor, in order for the model to accurately simulate a reactor's behavior. It should, however, be noted that when constructing models, it is typically a more fruitful procedure to start with simple models that will capture the correct qualitative behavior, and then refine the model by adding more complexity, until it captures the correct quantitative behavior.

Another issue making the search for a solution to the microkinetic differential equations for a catalytic process complex is that the allowable solutions correspond to coverages that are strictly in the interval from zero to one at all times, which strictly adds to one. If at any point a nonallowable step is taken, the equations will have a tendency to move toward a nonallowable solution. The differential equations are also extremely "stiff" in the sense that rate constants typically vary many orders of magnitude, and integration of the differential equations may lead to oscillations on disparate timescales, requiring the use of special solution techniques. Sometimes, the exact differences between very large and almost identical forward and backward rates are key to determine a stable solution, and just representing these with standard-accuracy arithmetic (e.g., typically 16 decimals) can be problematic. One may then have to use an arbitrary arithmetic package.

When we have more coverages varying in time, the temporal behavior of the reaction equations can become more complicated. When two independent coverages are present, the system can still move toward a steady state, but it can do this both in the same damped exponential fashion that it will in the single-coverage case, or it can move toward the stationary state in a damped oscillatory fashion. The steady state can also be unstable, and the system will then stay in undamped oscillatory motion. If there are three or more independent coverages, the corresponding system of nonlinear differential equations can additionally exhibit chaotic behavior.

Significant reduction in the complexity of solving the microkinetic equations can be achieved if we employ the approximation that the rate of change of all the coverages is zero, that is, *the steady-state approximation*:

$$\frac{\partial \theta_i}{\partial t} = 0 \quad \text{for all } i \tag{5.33}$$

Usually, this is a good approximation, but one should be careful, since it is not always guaranteed that there is only one set of coverages corresponding to a steady state. For cases where the coverages are time dependent due to an oscillation in rates, a steady-state approximation is still typically made. This relies on the (not proven) expectation for the time-averaged rate of the time-dependent microkinetic model to be similar to the rate of an unstable steady-state solution.

The steady-state approximation effectively turns the microkinetic model from a set of coupled nonlinear differential equations in time into a time-independent algebraic root-finding problem, which is simpler to solve and which can sometimes even be solved analytically. The rate corresponding to the steady-state solution is also continuous when the rate constants are varied continuously, which we shall see later can be useful for understanding trends. Whether steady state is a reasonable assumption is certainly debatable and is an issue that ought to be investigated in much further detail.

For the reactions (5.27, 5.28), the steady-state approximation amounts to

$$2k_1 \cdot p_{A_2} \cdot \left(1-\theta_A\right)^2 - 2k_{-1} \cdot \theta_A^2 - k_2 \cdot p_B \cdot \theta_A + k_{-2} \cdot p_{AB} \cdot \left(1-\theta_A\right) = 0, \quad (5.34)$$

which prescribes the steady solution through a second-order equation that can be solved for the steady-state coverage of species A. Although there will be two solutions of the second-order equation, only one of the two solutions will be positive and therefore correspond to a physically reasonable solution.

For many heterogeneously catalytic reactions, the overall reaction rate is typically determined by one specific elementary step being particularly "slow." The concept of a "slow" reaction needs to be clarified further. If we look at the microkinetic equations and invoke the steady-state approximation, it means that the net rates of the different reaction steps are approximately equal (except for some integer factors defined through the stoichiometry of the reaction). That is, the net production rates of the various types of adsorbates are zero. A "slow" reaction should therefore be understood as being so difficult for the system to carry out that it tends to be less equilibrated than all the other elementary steps. A useful concept is that of a "strongly rate-determining reaction step." By that, we shall mean a reaction step that is so difficult that all the other reaction steps are equilibrated, and they therefore have reached reversibilities of 1. The assumption that a specific elementary reaction step in a serial reaction is strongly rate determining (invoked in conjunction with the steady reaction conditions assumption, the steady-state assumptions, and the adsorbate–adsorbate noninteraction assumption) means that the kinetic equations become analytically solvable no matter how complex they were at the outset.

Assuming a strongly rate-determining step shall here generally refer to invoking all four aforementioned assumptions simultaneously, we make this choice, since the equations are generally only turned analytical by invoking all four assumptions simultaneously, and there are typically no other good reasons to assume a single rate-determining step, except to avoid solving the microkinetic model numerically and in order to give a closed expression for the rate. The analyticity is a general feature of simple equilibrium systems, and the assumption of a strongly rate-determining step in a serial reaction effectively establishes equilibrium for all reaction steps on either side of the rate-determining barrier.

To solve the reaction set (5.27, 5.28) in the strongly rate-determining approximation, we write the rates in terms of the following reversibilities:

$$R_1 = k_1 \cdot p_{A_2} \cdot \theta_*^2 \left(1-\gamma_1\right), \quad \gamma_1 = \frac{\theta_A^2}{p_{A_2} \theta_*^2} / K_1 \quad (5.35)$$

$$R_2 = k_2 \cdot p_B \cdot \theta_A \left(1-\gamma_2\right), \quad \gamma_2 = \frac{p_{AB} \theta_*}{p_B \theta_A} / K_2 \quad (5.36)$$

We note that a general feature for a serial catalytic cycle, which relies on reaction step i being carried out n_i times, the overall equilibrium constant is given by $K_{eq} = \prod_i K_i^{n_i}$ and the reversibility of the overall reaction is $\gamma = \prod_i \gamma_i^{n_i}$. For the reaction in question, we have $n_1 = 1$, and $n_2 = 2$, so the overall equilibrium constant is $K_{eq} = K_1 \cdot K_2^2$, and the catalytic reaction's overall reversibility is $\gamma_{eq} = \gamma_1 \cdot \gamma_2^2$.

Let us now assume that it is step one, which is strongly rate determining. This means that $\gamma_2 = 1$ and therefore $\gamma_1 = \gamma_{eq}$ (in general, we see that $\gamma_i = \gamma_{eq}^{1/n_i}$ if i is the rate-determining step). Since step 1 is rate determining, $\gamma_2 = 1 = \dfrac{p_{AB}\theta_*}{p_B\theta_A} / K_2$; lets us determine an expression for the fraction of A-covered sites, $\lambda_A = \theta_A / \theta_*$:

$$\lambda_A = p_{AB} \cdot p_B^{-1} \cdot K_2^{-1} \tag{5.37}$$

This type of expression is generally available for each adsorbate in a serial catalytic cycle when a strongly rate-determining step has been assumed. When each $\lambda_j = \theta_j / \theta_*$ has been determined, we can find the coverages of all surface species by using the site balance relation in the following form:

$$\theta_* + \sum_{j \neq *} \theta_j = 1 \tag{5.38}$$

in which we can take θ_* outside a parenthesis,

$$\theta_* \left(1 + \sum_{j \neq *} \lambda_j \right) = 1 \tag{5.39}$$

thereby expressing the coverage of free sites in known quantities:

$$\theta_* = \left(1 + \sum_{j \neq *} \lambda_j \right)^{-1} \tag{5.40}$$

from which all other coverages can be easily determined:

$$\theta_k = \lambda_k \left(1 + \sum_{j \neq *} \lambda_j \right)^{-1} \tag{5.41}$$

For the reaction in question, we thus get for the coverage of free sites

$$\theta_* = \left(1 + p_{AB} \cdot p_B^{-1} \cdot K_2^{-1} \right)^{-1} \tag{5.42}$$

and the coverage of A:

$$\theta_A = \left(p_{AB} \cdot p_B^{-1} \cdot K_2^{-1} \right) \left(1 + p_{AB} \cdot p_B^{-1} \cdot K_2^{-1} \right)^{-1} \tag{5.43}$$

which (when only one adsorbate is involved) can be written more compactly as

$$\theta_A = \left(1 + p_{AB}^{-1} \cdot p_B \cdot K_2\right)^{-1} \tag{5.44}$$

The rate of reaction is then found by taking the rate of the rate-determining step (since it has the only net rate that has not been assumed to be equal to zero) by inserting the appropriate coverages and using that $\gamma = \dfrac{p_{AB}^2}{p_{A_2} p_B^2} / K_{eq}$:

$$R = k_1 p_{A_2} \theta_*^2 \left(1 - \gamma\right) = k_1 p_{A_2} \left(1 + p_{AB} p_B^{-1} K_2^{-1}\right)^{-2} \left(1 - \frac{p_{AB}^2}{p_{A_2} p_B^2} / K_{eq}\right) \tag{5.45}$$

5.4   AMMONIA SYNTHESIS

We shall now take the ammonia synthesis reaction to exemplify the approach for an industrially relevant reaction. Ammonia synthesis has been discussed in Chapters 2 and 3, and the free energy diagrams have been shown in Figures 3.8 and 3.9 as a function of reaction temperature and pressure, respectively. As has been discussed in Chapter 2, synthesis of ammonia from N_2 and H_2 involves the following steps:

1. $N_2 + 2* \rightarrow 2N^*$
2. $H_2 + 2* \rightarrow 2H^*$
3. $N^* + H^* \rightarrow NH^* + *$
4. $NH^* + H^* \rightarrow NH_2^* + *$
5. $NH_2^* + H^* \rightarrow NH_3^* + *$
6. $NH_3^* \rightarrow NH_3 + *$

It is clear from Figures 3.8 and 3.9 that ammonia synthesis has to be carried out at high temperatures in order not to have the whole surface covered by intermediates. Similarly, a high pressure is required for the reaction to be exergonic at these temperatures. Figure 3.9 shows that at such conditions (high pressure (100 bar) and high temperature (700 K)), N_2 dissociation has by far the highest free energy barrier. We can therefore solve the kinetics of ammonia synthesis with the assumption of one rate-determining step (1) while treating all other steps as being equilibrated. We will do so using the mean field model under steady-state conditions as explained earlier.

The fact that steps (2) to (6) are equilibrated means that

$$\gamma_i = 1, \quad (i = 2 - 6) \tag{5.46}$$

Following the previous description, it follows that

$$K_2 p_{H_2} \theta_*^2 = \theta_H^2 \tag{5.47}$$

$$K_3 \theta_N \theta_H = \theta_{NH} \theta_* \tag{5.48}$$

$$K_4 \theta_{NH} \theta_H = \theta_{NH_2} \theta_* \tag{5.49}$$

$$K_5 \theta_{NH_2} \theta_H = \theta_{NH_3} \theta_* \tag{5.50}$$

$$K_6 \theta_{NH_3} = p_{NH_3} \theta_* \tag{5.51}$$

The first of these equations give the hydrogen coverage as

$$\theta_H = \sqrt{K_2 p_{H_2}} \, \theta_* \tag{5.52}$$

And the other equations can be arranged in the same way:

$$\theta_{NH_3} = \frac{p_{NH_3}}{K_6} \theta_* \tag{5.53}$$

$$\theta_{NH_2} = \frac{p_{NH_3}}{\sqrt{K_2 p_{H_2}} K_5 K_6} \theta_* \tag{5.54}$$

$$\theta_{NH} = \frac{p_{NH_3}}{K_2 p_{H_2} K_4 K_5 K_6} \theta_* \tag{5.55}$$

$$\theta_N = \frac{p_{NH_3}}{K_2^{3/2} p_{H_2}^{3/2} K_3 K_4 K_5 K_6} \theta_* \tag{5.56}$$

Combining with the site conservation rule (Equation (5.38), we can solve analytically to get the coverage of free sites as

$$\theta_* = \frac{1}{1 + \sqrt{K_2 p_{H_2}} + \dfrac{p_{NH_3}}{\sqrt{K_2 p_{H_2}} K_5 K_6} + \dfrac{p_{NH_3}}{K_2 p_{H_2} K_4 K_5 K_6} + \dfrac{p_{NH_3}}{K_2^{3/2} p_{H_2}^{3/2} K_3 K_4 K_5 K_6} + \dfrac{p_{NH_3}}{K_6}} \tag{5.57}$$

This allows us to write the rate of ammonia synthesis as

$$R = R_1 = k_1 p_{N_2} \theta_*^2 \left(1 - \gamma\right) \tag{5.58}$$

where

$$\gamma = \frac{p_{NH_3}^2}{K_{eq} p_{H_2}^3 p_{N_2}} \tag{5.59}$$

and

$$K_{eq} = K_1 K_2^3 K_3^2 K_4^2 K_5^2 K_6^2 \tag{5.60}$$

Equations (5.58) and (5.59) give the rate as a function of the partial pressures (p_{N_2}, p_{H_2}, p_{NH_3}) and temperature.

R_1 is the ammonia production per surface site (defined here as coverage 1, i.e., one monolayer of adsorbates) per second. This is also often referred to as the turnover frequency (TOF) of the reaction. The TOF of ammonia synthesis at a total pressure of 100 bar as a function of temperature on a Ru(0001) step as calculated in this simple model is shown in Figure 5.1.

Figure 5.1 shows that the simple model correctly captures that a temperature of the order 700 K is needed for the ammonia synthesis rate to be high enough for a reasonable TOF. Industrially, promoters are sometimes added to catalysts in order to increase their TOF significantly. This is also the case for ammonia synthesis where alkali and earth alkali promoters are employed. The role of these promoters is often to decrease the dissociation barrier of the rate-determining step. The issue of promoters will be discussed in Chapter 9.

Figure 5.2 shows the influence of the ratio between N_2 and H_2 on the TOF. Since N_2 dissociation is the rate-determining step, increasing the partial pressure of N_2 will decrease the free energy barrier of N_2 splitting and hence increase the TOF. In terms of conversion, however, maximum conversion is obtained with N_2:H_2 ratios of 1:3 since higher N_2 partial pressures shift the equilibrium toward the educts. Figure 5.2 is hence sensitive to the conversion level, for example, higher conversion shifts the maximum toward $p_{N_2} / (p_{N_2} + p_{H_2})$ close to 0.25. Industrially, ammonia synthesis is therefore conducted at ratios close to 1:3.

The model we used so far only provides a very approximate way of calculating the TOF. Yet, it captures important parts of the kinetics, and as we will see later, the real

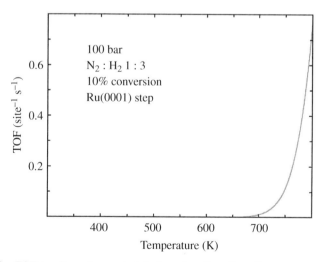

FIGURE 5.1 TOF per site and second plotted as a function of reaction temperature. Reaction conditions are as follows: $p = 100$ bar, $p_{N_2} : p_{H_2} = 1:3$, and conversion = 10%. Plot based on data of ammonia synthesis on the stepped Ru(0001) surface as obtained from CatApp and corrected for ZPE contributions (see also Figs. 3.8 and 3.9).

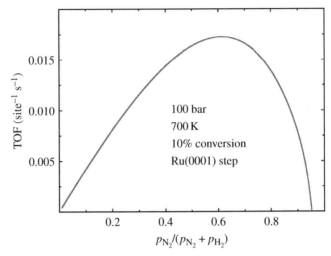

FIGURE 5.2 TOF per site and second plotted as a function of partial N_2 pressure. Reaction conditions are as follows: $p = 100$ bar, $T = 700$ K, and conversion $= 10\%$. Plot based on data of ammonia synthesis on the stepped Ru(0001) surface as obtained from CatApp and corrected for ZPE contributions (see also Figs. 3.8 and 3.9).

strength of simple models is that they can be used to look at trends in reactivity from one catalyst to the next.

To get a semiquantitative agreement with experimental rates on real catalysts, one has to extend the model by adding more complexity. So far, we have, for instance, only considered the reaction taking place on one kind of site of the catalyst. Interactions between different adsorbates on the surface have also been neglected in the mean field approach. We will show in the following a kinetic description of ammonia synthesis that includes the full complexity of interactions and reaction paths and is able to model the reaction under industrial conditions. Since the barrier for dissociation of N_2 depends on the environment around the dissociation site, we can extend Equation (5.58) by summation over different surface configurations (different neighbors around the dissociating N_2 molecule), each having different rate constants k_i (corresponding to the different activation energies):

$$R = \sum_i P_i k_i p_{N_2} \cdot (1 - \gamma) \tag{5.61}$$

Here, P_i is the probability of finding a configuration i. Coadsorption of species near the empty sites where N_2 dissociates will increase the dissociation barrier and hence decrease k_i. Equation (5.61) provides a way of including that complexity of the reaction. In addition, one needs to take care of the fact that N_2 dissociation at a step site proceeds with one N atom at the top of the step and another at the bottom; there are therefore two distinct sites that both have different binding energies and hence will have different coverages. The probability, P_i, must be calculated including the effect of interactions between all adsorbates.

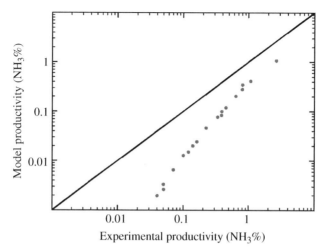

FIGURE 5.3 Comparison of the ammonia productivity from an extended microkinetic model with experimental results. Reaction conditions are as follows: 100 bar, $N_2:H_2 = 1:3$, flow range over 0.2 g catalyst between 40 and 267 ml/min, conversion range between 0 and 20 %, and temperature range between 320 and 440°C. Figure adapted from Honkala et al. (2005).

In order to compare the calculated rate to experiment, the only link is the number of active sites (steps) in the experimental setup. This can be estimated from electron microscopy, such as Figure 1.1. A direct comparison between the model and experiments at industrial conditions is shown in Figure 5.3.

While there is not quantitative agreement between theory and experiment, it is clear that the picture of the active site and the basic processes taking place during ammonia synthesis is quite good. This gives confidence that the picture developed here is a useful one.

REFERENCE

Honkala K, Hellman A, Remediakis IN, Logadottir A, Carlsson A, Dahl S, Christensen CH, Nørskov JK. Ammonia synthesis from first-principles calculations. Science 2005;307:555.

FURTHER READING

Bligaard T, Honkala K, Logadottir A, Nørskov JK, Dahl S, Jacobsen CJH. On the compensation effect in heterogeneous catalysis. J Phys Chem B 2003;107:9325.

Boudart M, Djéga-Mariadassou G. *Kinetics of Heterogeneous Catalytic Reactions.* Princeton University Press: Princeton; 1984.

Chorkendorff I, Niemantsverdriet JW. *Concepts of Modern Catalysis and Kinetics.* Wiley-VCH Verlag GmbH: Weinheim; 2003.

Dumesic JA, Rudd DF, Aparicio LM, Rekoske JE, Treviño AA. *The Microkinetics of Heterogeneous Catalysis*. Washington, DC: American Chemical Society; 1993.

Hansen EW, Neurock M. First-principles-based Monte Carlo simulation of ethylene hydrogenation kinetics on Pd. J Catal 2000;196:241–252.

Kandoi S, Greeley J, Sanchez-Castillo MA, Evans ST, Gokhale AA, Dumesic JA, Mavrikakis M. Prediction of experimental methanol decomposition rates on platinum from first principles. Top Catal 2006;37:17.

Linic S, Barteau MA. Construction of a reaction coordinate and a microkinetic model for ethylene epoxidation on silver from DFT calculations and surface science experiments. J Catal 2003;214:200–212.

Reuter K, Scheffler M. First-principles kinetic Monte Carlo simulations for heterogeneous catalysis: Application to the CO oxidation at RuO_2(110). Phys Rev B 2006;73:045433.

Stegelmann C, Andreasen A, Campbell CT. Degree of rate control: How much the energies of intermediates and transition states control rates. J Am Chem Soc 2009;131:8077.

Temel B, Meskine H, Reuter K, Scheffler M, Metiu H. Does phenomenological kinetics provide an adequate description of heterogeneous catalytic reactions? J Chem Phys 2007; 126:204711.

van Santen RA, Niemantsverdriet JW. *Chemical Kinetics and Catalysis*. New York: Plenum Press; 1995.

Zhadanov VP, Kasemo B. Kinetics of rapid heterogeneous reactions on the nanometer scale. J Catal 1997;170:377.

6

ENERGY TRENDS IN CATALYSIS

Obtaining a full kinetic description of a surface chemical reaction involves measuring or calculating the binding energy of each intermediate in the reaction together with the energy barriers needed to go from one state to the next. In order to understand trends in catalytic activity, one would need this information for a number of different catalytic materials with different surface structures. In the following, we describe a set of tools that can be used to reduce the parameter space that one needs to cover to get a qualitative description of the reaction energetics involved in a given process. The end result is a mapping of the full potential energy diagram onto a limited set of parameters, which we call descriptors. Understanding trends in reactivity then becomes a question of understanding trends in a limited set of parameters.

In this chapter, we will discuss how adsorption energies and activation energies for surface chemical processes are often correlated. As discussed in Chapter 2, there are two distinct types of surface–adsorbate interactions: physisorption and chemisorption (see Fig. 6.1). We will treat different kinds of correlations in the two types of interactions separately in the following.

6.1 ENERGY CORRELATIONS FOR PHYSISORBED SYSTEMS

The strength of the van der Waals interactions responsible for physisorption depends strongly on the spatial extent of the adsorbate, and for comparable systems, these should scale well when normalized to their active components.

Fundamental Concepts in Heterogeneous Catalysis, First Edition. Jens K. Nørskov,
Felix Studt, Frank Abild-Pedersen and Thomas Bligaard.
© 2014 John Wiley & Sons, Inc. Published 2014 by John Wiley & Sons, Inc.

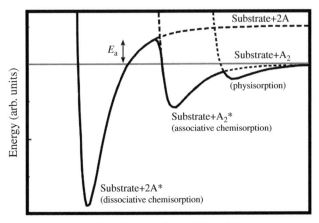

FIGURE 6.1 Schematic energy profile for the adsorption of some diatomic molecule A_2 on some substrate as a function of a specific reaction coordinate, which could be the distance between the substrate surface and the center of mass of the molecule.

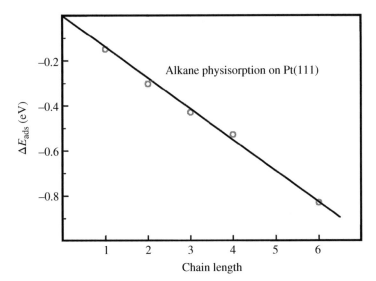

FIGURE 6.2 Experimental data for physisorption of linear alkanes on Pt(111). Adapted from Tait et al. (2006).

In Figure 6.2, experimental results for alkane physisorption on Pt(111) are plotted as a function of chain length showing a linear trend. This suggests that each $-CH_x-$ unit within the alkane chain has a unique contribution of the order 0.15 eV to the bond strength and that the overall binding energy is given approximately by the sum of these contributions. This greatly simplifies the description of physisorption systems and contributions from van der Waals interactions to bonding (including stability of transition states) from larger molecules where such effects can dominate.

6.2 CHEMISORPTION ENERGY SCALING RELATIONS

We next consider situations where there is considerable interaction between adsorbate and adsorbent. The strength of the interaction is such that the electronic structure of the adsorbate changes significantly and a chemical bond is formed. In Chapters 8 and 12, we will see how to understand the bonding between adsorbates and surfaces in more detail.

It has been found very generally that adsorption energies of different surface intermediates that bind to the surface through the same atom(s) scale with each other. In Figure 6.3, we show examples of scaling relations for CH_n species ($n = 1, 2, 3$). What is shown is the adsorption energy of CH_n on a number of metals and for two different surface structures plotted as a function of the adsorption energy of atomic C. For a given adsorbate and surface structure, the relationship is to a good approximation linear

$$\Delta E_{CH_n} = \gamma_s(n)\Delta E_C + \xi \tag{6.1}$$

and, interestingly, to a good approximation $\gamma_s(n) = (4-n)/4$. Similar plots for NH_n and OH_n species give $\gamma_s(n) = (3-n)/3$ and $\gamma_s(n) = (2-n)/2$, respectively.

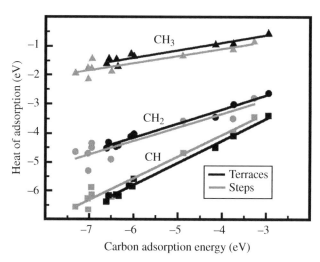

FIGURE 6.3 DFT-calculated adsorption energies of CH_x species plotted against the adsorption energy of C for a number of different transition metals. The black and gray symbols indicate close-packed and stepped surfaces, respectively. The lines show the best fits to the points. Adapted from Abild-Pedersen et al. (2007).

The results of Figure 6.3 suggest a simple valency rule for the scaling slope. If A has a maximum valence of N and $n \leq N$ of these bonds is saturated by bonding to other atoms, then the scaling slope is given by:

$$\gamma_s(n) = \frac{N-n}{N} \tag{6.2}$$

The simplest way to understand this is to say that the N levels are degenerate, separable, and linearly independent such that each contributes with the same amount to the total binding energy when A couples to a surface. Saturating one of these energy levels by bonding to H or another atom will remove it from the equation and hence reduce the number of bonds the element can make with the surface.

We note that the scaling parameter is independent of the surface. The cutoff, ξ, in Equation (6.1), on the other hand, depends on the surface structure. Hence, in order to obtain information about how well a given structure binds an adsorbate, given that one knows how all the base elements (C, O, N, S, etc.) bind, one needs to do only a single value measurement or calculation of the adsorption energy, and the rest can be scaled from the binding of the base elements. This enables us to write a simple expression for the reaction energy of an elementary step as

$$\Delta E = \sum_{i=1}^{N} \left(\Delta\gamma_i \Delta E^{A_i} \right) + \Delta\xi \tag{6.3}$$

Here, the sum is over all atoms, i, forming bonds to the surface, $\Delta\gamma_i$ is the change in the scaling slope or valency parameter during the reaction, ΔE^{A_i} denotes the binding energy of the base elements relevant for the reaction, and $\Delta\xi$ is a constant that one has to measure or calculate for a single system.

In Figure 6.4, it is shown how well the reaction energies are described using the model compared to the full DFT calculations.

For any unsaturated hydrocarbon with n carbon atoms, we would expect that

$$\gamma_s = \sum_{j=1}^{n} \left(1 - \sum_{i=1}^{k} \frac{1}{4} \right) \tag{6.4}$$

where k is the number of saturated bonds per carbon atom. In Figure 6.5, we show adsorption energies of CH_x–CH_2 where $x \in \{0,1,2\}$ on close-packed (111) and stepped (211) surfaces as a function of the adsorption energy of atomic carbon. The scaling relations define two regions: one region where the metal surface is reactive enough to alter the intrinsic C–C bond and another region where the surface is too noble. The theoretical slopes given by Equation (6.4) have been used to obtain the best-fit lines in Figure 6.5.

It is evident from the earlier discussion that scaling among adsorption energies should not be limited to transition metal surfaces. In fact, even for metal-terminated surfaces of more complex systems like transition metal compounds (oxides, nitrides, sulfides, and carbides), where there is mixed covalent, ionic bonding between the

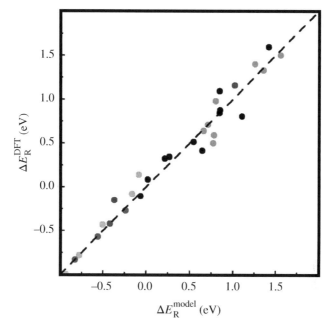

FIGURE 6.4 Graph shows DFT-calculated reaction energies against model reaction estimated using Equation (6.3). The metal-/structure-dependent parameter has been calculated for Pt(111). Reactions involved are dehydrogenation of CH_3OH (black), C_3H_5 (red), CH_3SH (green), cysteine (blue), and ethylene (magenta). Adapted from Abild-Pedersen et al. (2007). (*See insert for color representation of the figure.*)

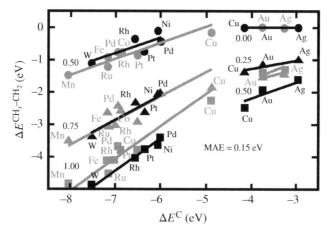

FIGURE 6.5 Calculated adsorption energies of $CH_x–CH_2$ ($x = 0, 1, 2$) intermediates as a function of adsorption energies of atomic carbon (circles $x = 2$; triangles $x = 1$; squares $x = 0$). The data points shown in black represent the close-packed surfaces, and the data points in gray are for stepped surfaces. The solid lines represent best fits to the data points given a theoretical slope as defined in Equation (6.4). The mean absolute error (MAE) is shown for all data. Adapted from Jones et al. (2011).

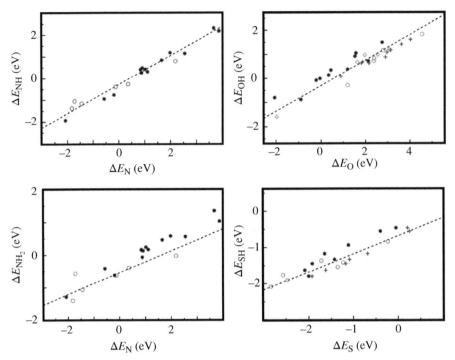

FIGURE 6.6 Figures show NH$_x$ versus N scaling on transition metals and metal nitrides, OH versus O scaling on transition metals and metal oxides, and SH versus S scaling relations on transition metals and metal sulfides. Data points involve stepped transition metal surfaces (black), close-packed transition metal surfaces (blue), and the nitride, oxide, and sulfide surfaces (red). The dashed line shows the best fit to the points using the theoretical slope. Adapted from Fernandez et al. (2008). (*See insert for color representation of the figure.*)

surface cations and anions, there is scaling between electronically similar adsorbates. In Figure 6.6, we show such relations for oxide, nitride, and sulfide surfaces.

6.3   TRANSITION STATE ENERGY SCALING RELATIONS IN HETEROGENEOUS CATALYSIS

Given that the variations in adsorption energies and transition state energies are governed by the same basic physics, it is not surprising that transition state energies also correlate with adsorption energies. As for the relationships between adsorption energies, scaling between adsorption energies and transition state energies is extremely important in building an understanding of heterogeneous catalysis, and at a more fundamental level, they provide guidance in building kinetic models to understand trends in catalytic activity.

Let us begin by defining a more general relationship between transition state energies and adsorption energies. Let E^{TS} be a set of energies describing the energy

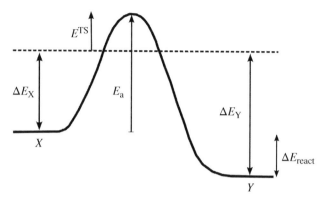

FIGURE 6.7 Potential energy surface showing schematically the relevant energies needed to describe the jump between two minima X and Y. E_a and E^{TS} both describe the first-order saddle point between X and Y relative to the energy of X and the energy reference defined by the dashed line, respectively. ΔE_{react} is the reaction energy defined by the difference in energy of state Y and state X, and ΔE_X and ΔE_Y are energies relative to the reference energy of state X and state Y, respectively.

needed to move between two minima on the potential energy surface for a set of different catalysts. Furthermore, let ΔE_i be a set of adsorption energies, relevant for the process of moving between the two minima. We can now define a functional form of $E^{TS}(\Delta E_i)$, which is a map from the space of adsorption energies to the space of transition state energies.

To first order in ΔE_i, E^{TS} will be given as a linear combination of ΔE_i:

$$E^{TS} = \sum_i \gamma_i \Delta E_i + \xi \qquad (6.5)$$

The set of functions defined in Equation (6.5) constitute a class of linear relations—the linear transition state energy scaling relations. Note that we can always restrict the variation of adsorption energies to be small enough for the scaling to be linear, but for a broad enough range of energies, nonlinear scaling is always found.

The transition state scaling relations imply scaling relations for the activation energy of a surface chemical reaction (see Fig. 6.7). Let X and Y define two minima on a potential energy surface; if both E^{TS} and ΔE_X scale with a set of adsorption energies, then E_a will as well.

Linear correlations between activation (free) energies and reaction (free) energies is a well-established approach in the understanding of trends in chemical reactions that dates back to Brønsted in 1928 and Evans and Polanyi 10 years later. Such BEP relations have for a long time been assumed to hold in heterogeneous catalysis. BEP relations can be considered a subset of the relations defined in Equation (6.5). As illustrated in Figure 6.7, the activation energy of an elementary process is the energy difference between a transition state energy and the energy of an intermediate.

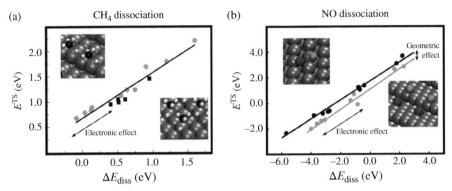

FIGURE 6.8 (a) Linear transition state energy relationship for dehydrogenation of methane over a number of fcc (211) (solid light gray circles) and (111) (solid black squares) transition metal surfaces. The stepped surfaces show a slightly higher dissociation barrier than the (111) surfaces at a given dissociative chemisorption energy ΔE_{diss}, but the electronic effect is much larger than the geometrical effect. (b) The linear transition state energy relationship for NO dissociation over a number of stepped (solid light gray circles) and close-packed (solid black circles) surfaces as a function of the dissociation energy. The line for open surfaces lies significantly below that of the close-packed surfaces (on the order of 0.7 eV). At given reactivity of the surface, NO thus prefers dissociating over the undercoordinated sites at the steps. Here, the geometric effect is larger than the electronic effect. Adapted from Nørskov et al. (2008) and Falsig et al. (2013).

Similarly, reaction energies are differences between energies of two intermediates. If a BEP relation exists, then there will also be a transition state scaling relation. There can, however, be many transition state scaling relations that are not covered by a BEP relation. By restricting the independent variable to the reaction energy in a BEP relation, one does not derive the full potential of the scaling relations.

It is only with the advent of sufficiently accurate electronic structure calculations that it has become possible to obtain transition state energies and adsorption energies over a wide enough energy range to establish transition state scaling relations with sufficient statistics. Figure 6.8 shows examples of linear transition state scaling relations.

The slopes of these relations depend on the reaction studied. For dissociative adsorption processes involving simple diatomic molecules, the slope of the transition state energy as function of the dissociative chemisorption energy is often close to 1. This implies that the electronic structure of the transition state is similar to that of the final state, and hence, it is indicative of a late transition state. This behavior can be observed directly in the transition state structures for NO dissociation, as is shown in Figure 6.8b.

The transition state scaling relations provide a rigorous way of defining the effect of surface structure on reactivity. Such a definition is not simple in general since a change in surface structure also changes the surface electronic structure, and it is difficult to distinguish between purely geometrical and the surface structure-related electronic structure effects. The transition state scaling relations allow one to look at differences in transition state energy for a fixed adsorption energy. To the extent that the adsorption energy defines the electronic structure effects, then a shift in a scaling line is a direct measure of effects linked to the local geometry of the active site. We note that in principle there is a different line for every surface geometry, so one should think of a family of transition state scaling lines. For NO dissociation, a large number of geometries have been investigated, and the lines for the close-packed and the stepped surface shown in Figure 6.8b basically define the upper and lower bound to the scaling lines. A close-packed surface and a stepped surface therefore represent two limits to reactivity for a given catalyst material and form a good test ground for understanding structural effects in catalysis.

A guideline for identifying surface geometries with low-lying lines is to find surfaces where the two fragments of the reaction both can be stabilized without too many "shared" metal atoms. In Chapters 8 and 12, we will discuss electronic structure effects in more detail.

6.4 UNIVERSALITY OF TRANSITION STATE SCALING RELATIONS

It turns out that if one compares dissociation of a number of similar molecules, their transition state energy scale with the dissociative chemisorption energy in much the same way (see Fig. 6.9). This is a remarkable result indicating that the nature of the relationship between the final state and the transition state for dissociation of these molecules is quite similar. In fact, for the large number of systems considered in Figure 6.9, essentially, all transition states look the same for a given surface structure.

We note that the geometrical effect discussed for NO dissociation holds for all the adsorbates considered in Figure 6.9. This means that CO, N_2, and O_2 dissociation should also be much faster at steps than at the most close-packed surface. The linear transition state scaling relations provide an adsorbate-independent way of estimating activation energies based on the adsorption energies of the product species. As we shall see in Chapter 7, these relationships can be used to obtain information about how good a material is as a catalyst for a given reaction. The universal relation for the close-packed surfaces shown in Figure 6.9 is found to be

$$E^{TS} = (0.90 \pm 0.04) \cdot \Delta E_{diss} + (2.07 \pm 0.07) \, eV$$

and the relation for the stepped surfaces is

$$E^{TS} = (0.87 \pm 0.05) \cdot \Delta E_{diss} + (1.34 \pm 0.09) \, eV$$

The slopes of the relations are very similar and close to one, showing that the transition states of the reactants considered indeed are very final state like. The intercepts are different, and this difference identifies the structure dependence of the relations showing that for a given value of the dissociative chemisorption energy, over a range of relevant energies, the stepped surfaces have barriers that are much less than on the close-packed surfaces.

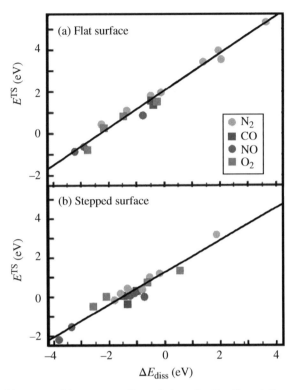

FIGURE 6.9 Linear transition state scaling relations for the dissociation of a number of simple diatomic molecules. It is clear from the plots that for a given surface geometry, all the data cluster around the same "universal" line. Adapted from Nørskov et al. (2002). (*See insert for color representation of the figure.*)

These relationships turn out to be more general, thus including other classes of reactions and different kinds of surface terminations. In Figure 6.10, we show that to a reasonable approximation, calculated transition state energies and dissociative chemisorption energies on a series of coupling reactions involving C–C, C–N, C–O, N–O, and O–O species follow a universal relation. In Figure 6.11, we see the same behavior for diatomic molecules but now on a series of different transition metal oxide surfaces in the rutile structure.

We note that there are exceptions to these scaling relations, especially when molecules with weak interatomic bonds are considered such as in the dissociation of H_2. However, all these deviations can be understood in terms of general models within electronic band structure theory.

It is important to stress that even though deviations from the linear behavior are seen, these correlations are sufficient to describe trends in reactivity, as we shall see in the next chapter. We also note that a higher accuracy can be obtained if one considers one specific reaction only.

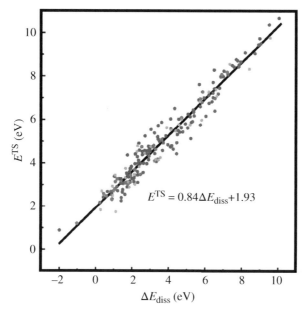

FIGURE 6.10 Universal relation between a large number of coupling reactions on transition metal surfaces. The reactions involved are C–C coupling reactions (blue), C–O coupling reactions (red), C–N coupling reactions (orange), N–N coupling reactions (purple), N–O coupling reactions (green), and O–O coupling reactions (turquoise). Adapted from Wang et al. (2011). (*See insert for color representation of the figure.*)

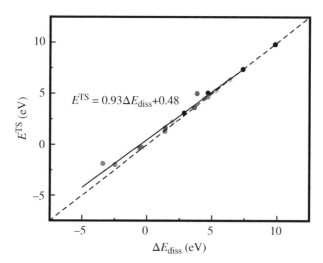

FIGURE 6.11 Universal relation between calculated transition state energies and dissociative chemisorption energies for diatomic systems on (110) transition metal oxide surfaces in the rutile structure. N_2 splitting reactions (black), O_2 splitting reactions (red), CO splitting reactions (orange), and NO splitting reactions (blue) are plotted. Adapted from Vojvodic et al. (2011). (*See insert for color representation of the figure.*)

REFERENCES

Abild-Pedersen F, Greeley I, Studt F, Rossmeisl I, Munter TR, Moses PG, Skulason E, Bligaard T, Nørskov JK. Scaling properties of adsorption energies for hydrogen-containing molecules on transition-metal surfaces. Phys Rev Lett 2007;99:016105.

Falsig H, Shen J, Khan TS, Guo W, Jones G, Dahl S, Bligaard T. On the structure sensitivity of direct NO decomposition over low-index transition metal facets. Top Catal 2013;57: 80–88.

Fernandez EM, Moses PG, Toftelund A, Hansen HA, Martinez JI, Abild-Pedersen F, Kleis J, Hinnemann B, Rossmeisl J, Bligaard T, Nørskov JK. Scaling relationships for adsorption energies on transition metal oxide, sulfide, and nitride surfaces. Angew Chem Int Ed 2008; 47:4683–4686.

Jones G, Studt F, Abild-Pedersen F, Nørskov JK, Bligaard T. Scaling relationships for adsorption energies of C2 hydrocarbons on transition metal surfaces. Chem Eng Sci 2011;66:6318.

Nørskov JK, Bligaard T, Logadottir A, Bahn S, Hansen LB, Bollinger M, Bengaard H, Hammer B, Sljivancanin Z, Mavrikakis M, Xu Y, Dahl S, Jacobsen CJH. Universality in heterogeneous catalysis. J Catal 2002;209:275–278.

Nørskov JK, Bligaard T, Hvolbæk B, Abild-Pedersen F, Chorkendorff I, Christensen CH. The nature of the active site in heterogeneous metal catalysis. Chem Soc Rev 2008;37: 2163–2171.

Tait SL, Dohnalek Z, Campbell CT, Kay BD. n-alkanes on Pt(111) and on C(0001)/Pt(111): chain length dependence of kinetic desorption parameters. J Chem Phys 2006;125:234308.

Vojvodic A, Calle-Vallejo F, Guo W, Wang S, Toftelund A, Studt F, Shen J, Man IC, Rossmeisl J, Bligaard T, Nørskov JK, Abild-Pedersen F. On the behavior of Brønsted-Evans-Polanyi relations for transition metal oxides. J Chem Phys 2011;134:244509.

Wang SG, Temel B, Shen JA, Jones G, Grabow LC, Studt F, Bligaard T, Abild-Pedersen F, Christensen CH, Nørskov JK. Universal Brønsted-Evans-Polanyi relations for C–C, C–O, C–N, N–O, N–N, and O–O dissociation reactions. Catal Lett 2011;141:370–373.

FURTHER READING

Abild-Pedersen F, Greeley J, Studt F, Rossmeisl J, Munter TR, Moses PG, Skulason E, Bligaard T, Nørskov JK. Scaling properties of adsorption energies for hydrogen-containing molecules on transition-metal surfaces. Phys Rev Lett 2007;99:016105.

Michaelides A, Liu ZP, Zhang CJ, Alavi A, King DA, Hu P. Identification of general linear relationships between activation energies and enthalpy changes for dissociation reactions at surfaces. J Am Chem Soc 2003;125:3704–3705.

Nørskov JK, Bligaard T, Logadottir A, Bahn S, Hansen LB, Bollinger M, Bengaard H, Hammer B, Sljivancanin Z, Mavrikakis M, Xu Y, Dahl S, Jacobsen CJH. Universality in heterogeneous catalysis. J Catal 2002;209:275–278.

7

ACTIVITY AND SELECTIVITY MAPS

As described in detail in Chapter 5, elaborate kinetic methods are available to provide a detailed description of the rate of a given heterogeneous reaction. In this chapter, we shall focus on the more general description of trends in catalysis. Mean-field microkinetic models are in many cases adequate for semiquantitatively describing the reaction rate and have some distinct advantages when studying trends, since the introduction of a few additional assumptions (such as inclusion of a rate-determining reaction step and the steady-state approximation) will often result in the model becoming entirely analytical.

The microkinetic models in this section are built upon scaling relations of the type described in Chapter 6. It will be shown that an underlying scaling relation in general leads to the existence of what we call an *activity map*, which is a map of the catalytic activity as a function of a few descriptors. In many cases, such a map shows a single maximum for a certain set of descriptor values, and it is often also called a *volcano relation*. In a similar way, we will introduce selectivity maps showing selectivity as a function of the descriptors.

7.1 DISSOCIATION RATE-DETERMINED MODEL

We start by considering the simplest possible surface-catalyzed reaction treated in some detail in Chapter 5:

$$A_2 + 2B \rightarrow 2AB \tag{7.1}$$

Fundamental Concepts in Heterogeneous Catalysis, First Edition. Jens K. Nørskov,
Felix Studt, Frank Abild-Pedersen and Thomas Bligaard.
© 2014 John Wiley & Sons, Inc. Published 2014 by John Wiley & Sons, Inc.

The reaction scheme of elementary steps can be written as

$$A_2 + 2* \rightarrow 2A* \tag{7.2}$$

$$A* + B \rightarrow AB + * \tag{7.3}$$

where an asterisk represents an active surface site. A schematic energy diagram for this reaction is shown in Figure 7.1.

Let us assume that A_2 dissociation is rate limiting. The turnover frequency (TOF) of the reaction is then (see also Eq. 5.35)

$$r(T,p) = k_1 p_{A_2} \theta_*^2 (1-\gamma) \tag{7.4}$$

Here, k_1 is the temperature-dependent rate constant for the forward reaction in Equation (7.2), which is assumed to follow an Arrhenius expression $k_1 = \dfrac{k_B T}{h} e^{\Delta S_a / k_B} e^{-E_a / k_B T}$. γ is the overall gas-phase approach to equilibrium, which can be written as

$$\gamma = \frac{p_{AB}^2}{K_{eq} p_{A_2} p_B^2} \tag{7.5}$$

where K_{eq} is the equilibrium constant for the overall reaction, p_B is the pressure of reactant B, and p_{AB} is the pressure of the product, AB. The coverage of free sites can be determined analytically as

$$\theta_* = \frac{1}{1 + \dfrac{\theta_A}{\theta_*}} = \frac{1}{1 + \dfrac{p_{AB}}{K_2 p_B}} = \frac{1}{1 + \sqrt{K_1 p_{A_2} \gamma}} \tag{7.6}$$

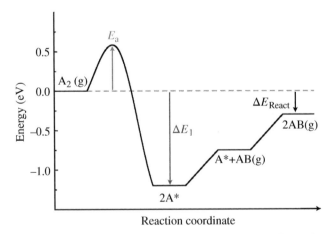

FIGURE 7.1 Potential energy diagram for the reaction $A_2 + 2B \rightarrow 2AB$ where A_2 adsorbs dissociatively on the surface and B reacts without prior adsorption with adsorbed A.

where $K_1 = e^{-\Delta G_1^\circ / k_B T}$ is the equilibrium constant for the reaction step in Equation (7.2) with the standard Gibbs free energy of $\Delta G_1^\circ = \Delta E_1 - T\Delta S_1$. Under the aforementioned assumptions, it is possible to obtain an analytical expression for the TOF. The dissociative chemisorption energy ΔE_1 is the quantity that determines the number of free sites. Hence, very reactive surfaces (surfaces with very negative ΔE_1) will poison the reaction in the sense that there will be very few sites for A_2 to dissociate. The TOF will thus decrease when $\Delta E_1 \to -\infty$.

For a given surface, the transition state scaling relation discussed in Chapter 6 relates the activation energy of step 1, E_a, to the dissociative chemisorption energy, ΔE_1: $E_a = \gamma \Delta E_1 + \xi$. The activation barrier will be large on very noble surfaces (ΔE_1 numerically small and negative or even positive), and as a consequence, the TOF will decrease as $\Delta E_1 \to \infty$. In the intermediate ΔE_1 range, the TOF passes through a maximum.

We note that there are two parameters describing the catalyst, ΔE_1 and E_a, and because of the transition state scaling relation, there is only one independent variable, which we choose to be ΔE_1. The scaling relation means that there is a single descriptor of reactivity, ΔE_1. We will show later in this chapter how the scaling relations allow the identification of a few descriptors of reactivity even for more complicated reactions.

To illustrate how to analytically determine the activity map for the simple model reaction, we will make some further (rather arbitrary but reasonable) assumptions:

- The reaction energy of Reaction (7.1) is −0.3 eV (see also Fig. 7.1).
- The entropy of gas-phase A_2 and AB is 0.002 eV/K. The entropy for gas-phase B is 0.0015 eV/K (see also entropies for gas-phase species in Chapter 3).
- The entropy of element A adsorbed on the surface is assumed to be negligible. This is not generally true, but it is often a sufficiently good approximation (see Chapter 3).
- The transition state for the dissociative chemisorption and for desorption is strongly constrained. This assumption allows us to set the transition state entropy to zero, and the prefactors in the Arrhenius expressions for the rate constants of desorption of AB and redesorption of A_2 thus become $\dfrac{k_B T}{h}$ (see Chapter 4).
- Dissociation of A_2 follows a transition state scaling line with the slope of 0.87 and the intercept of 1.34 eV, which is what is found for the dissociation of diatomic molecules on the 211 steps of transition metal surfaces (see Chapter 6).

As shown in Equation (7.4), the TOF depends on the product of k_1 and θ_*^2. The dependencies of k_1 and θ_* on ΔE_1 are shown in Figure 7.2 together with the overall TOF r. While the number of free sites, θ_*, decreases with higher ΔE_1 due to poisoning of the surface by species A*, k_1 increases steadily until the barrier for A_2 dissociation becomes zero and $k_1 \sim \dfrac{k_B T}{h} \sim 10^{13}$. The maximum TOF is obtained

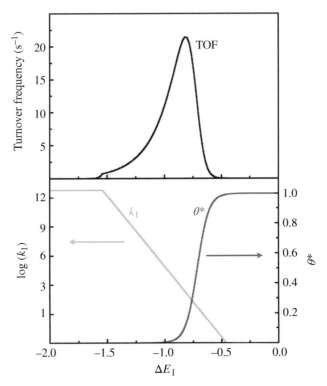

FIGURE 7.2 *Bottom*: coverage of free sites, θ_*, and logarithm of the forward rate, k_1, plotted as a function of ΔE_1. *Top*: TOF plotted as a function of ΔE_1. Reaction conditions are $T = 300$ K, $p = 30$ bar, $p_{A_2} : p_B = 1:2$, and conversion = 10%.

when θ_* is around 0.5 (see Fig. 7.2). This can be viewed as a general rule and the top of highest activity is located at approximately this position for many heterogeneously catalyzed reactions.

The result in Figure 7.2 illustrates the Sabatier principle, which states that the catalytic activity for a given reaction follows a volcano-shaped curve, because only an intermediate binding of intermediates on the surface of a catalyst will give a reasonably active catalyst. We can understand the origin of this principle based on the treatment earlier. The underlying, implicit assumption is that there is a monotonic relationship between the rate of activating the reactants and the rate of formation of the products. This is exactly what the transition state scaling relations provide. The Sabatier principle was formulated almost 100 years ago and provided an extremely valuable qualitative way of understanding why there is an optimum catalyst. The analysis outlined earlier provides three additional insights. First, it provides a way to identify which "binding" needs to be "intermediate"; that is, it provides a systematic way of identifying which adsorption energies are descriptors (ΔE_1 for the simple problem earlier) for a given reaction. These are the properties that need to be optimal to give the highest rate. The tool for identifying descriptors is the concept of scaling relations. The second, crucial new component is that having identified the descriptors we can quantify them. That means that we can go beyond saying that "there is an optimum value" to identifying which is the optimum value. This makes the analysis predictive and useful in identifying leads for new catalysts. Finally, the analysis allows an understanding of how the optimum depends on reaction conditions. In the following, we first use the simple generic reaction to discuss some general principles for how the optimum catalyst depends on reaction conditions. We then introduce a simplified analysis tool, and finally, we apply the approach to examples of real catalyst reactions.

7.2 VARIATIONS IN THE ACTIVITY MAXIMUM WITH REACTION CONDITIONS

We will now show how the activity map changes when the approach to equilibrium, the temperature, and the pressure of the reaction are varied independently. Various reactions that proceed via the dissociation of diatomic molecules can have drastically different overall reaction energies. It is therefore often very useful to write the microkinetics in terms of the approach to equilibrium instead of the product pressure and reaction energy. Reactions with different reaction energies all follow the same microkinetic model for a given approach to equilibrium. However, for the various reactions, a given approach to equilibrium will correspond to very different product pressures and hence a different reaction conversion. The activity maps obtained for different approaches to equilibrium for the microkinetic model discussed earlier with dissociation as rate determining are shown in Figure 7.3. The TOFs are plotted as a function of the dissociative adsorption energy, ΔE_1. The activity curves increase

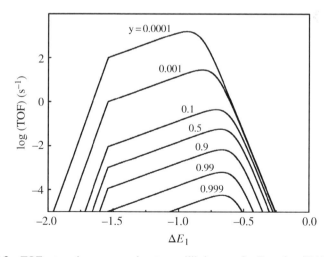

FIGURE 7.3 TOFs at various approaches to equilibrium, γ, for Reaction (7.1) plotted as a function of ΔE_1. Reaction conditions are otherwise $T = 300$ K, $p = 30$ bar, and $p_{A_2} : p_B = 1:2$.

in magnitude when the approach to equilibrium becomes smaller with their top shifting to stronger bonding (more negative ΔE_1). This is especially true when the approach to equilibrium has a very low value ($\gamma < 0.001$). In practice, this often indicates that one does not necessarily want to use the same catalyst at the front end of a tubular reactor, where conversion levels are still low, as at the back end, where conversion levels are high and γ will be closer to one.

Ammonia synthesis is an example of a process where activation of the reactant (N_2) is rate determining for the most interesting catalysts. In that process, iron is the optimal catalyst far from equilibrium, but once the last part of the reactor bed is reached and $\gamma \sim 1$, ruthenium, which is more noble (and more expensive), has been used to replace iron. Since $K_{eq} = e^{\frac{-\Delta G^\circ}{k_B T}}$, the more exothermic a reaction is, the larger K_{eq} becomes. For a given conversion, the approach to equilibrium, $\gamma = \dfrac{p_{AB}^2}{K_{eq} p_{A_2} p_B^2}$, therefore becomes smaller. This means that the maximum of the activity curve shifts to more negative ΔE_1 for more exothermic reactions.

We will now use the same microkinetic model to investigate how the activity map changes with temperature. In Figure 7.4, the dependence of the TOF on the temperature is shown. For high temperatures, the optimal catalyst moves toward more reactive surfaces, whereas more noble catalysts are closer to the optimum at lower temperature. The decisive factor for this is the availability of free sites on the surface where A_2 can dissociate rather than the activation of A_2 itself. Higher reaction temperatures are driving products away from the surface (due to their increased entropy contribution in the gas phase) and thus provide more free sites.

Figure 7.5 shows how the TOF depends on the pressure of A_2, the most important reactant in the process we are treating here. The activity curve does increase with

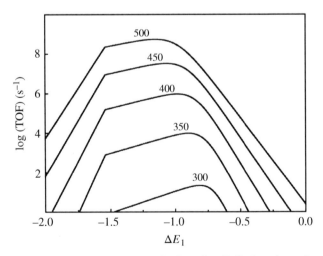

FIGURE 7.4 TOFs at various temperatures for Reaction (7.1) plotted as a function of ΔE_1. Reaction conditions are otherwise $p = 30$ bar, $p_{A_2} : p_B = 1:2$, and conversion $= 10\%$.

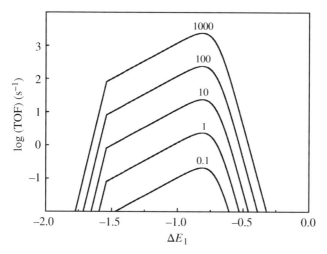

FIGURE 7.5 TOFs at various pressures of A_2 for Reaction (7.1) plotted as a function of ΔE_1. Reaction conditions are otherwise $p = 30$ bar, $p_{A_2} : p_B = 1{:}2$, and conversion $= 10\%$.

increasing pressure of A_2. However, its position varies much less with pressure than it varies due to changes in the approach to equilibrium and the temperature.

7.3 SABATIER ANALYSIS

We now introduce a method that provides the simplest possible conceptual framework for analyzing microkinetic models of heterogeneous reactions, the so-called Sabatier analysis. We call it so because it brings out the qualitative reasoning behind the Sabatier principle in a quantitative form.

Consider again Reaction (7.1), but let us now relax the assumption that the activation of A_2 is rate determining. The approach to equilibrium for the full reaction is shown in Equation (7.5), and we can write the equilibrium constant in terms of equilibrium constants for the two elementary steps:

$$K_{eq} = K_1 K_2^2 \tag{7.7}$$

and hence

$$\gamma = \gamma_1 \gamma_2^2 \tag{7.8}$$

We will now focus on the net reactions proceeding in the forward direction so that

$$0 \leq \gamma \leq 1 \tag{7.9}$$

$$0 \leq \gamma_1 \leq 1 \tag{7.10}$$

$$0 \leq \gamma_2 \leq 1 \tag{7.11}$$

and using Equation (7.8), this means that

$$0 \leq \gamma \leq \gamma_1 \leq 1 \tag{7.12}$$

$$0 \leq \gamma \leq \gamma_2^2 \leq 1 \tag{7.13}$$

We will start by analyzing the reaction on the surface of a catalyst that bonds intermediates too strongly (to the left of the maximum in the activity map, Fig. 7.2). The surface coverage will be high ($\theta_A \approx 1$) and desorption of AB will be the rate-determining step. This means that the first reaction step (dissociative adsorption of A_2, Eq. 7.2) is in equilibrium and hence that $\gamma_1 \approx 1$ and $\gamma_2 \approx \sqrt{\gamma}$. The TOF, r_{tot}, can now be approximated via the second reaction step, r_2:

$$r_{tot} = r_2 = k_2 p_B \left(1 - \sqrt{\gamma}\right) \tag{7.14}$$

For too noble surfaces (to the right of the maximum in the activity map, Fig. 7.2), where dissociation of A_2 is rate determining, the coverage of free sites will be approximately 1 ($\theta_* \approx 1$), and we have the second step in equilibrium so that $\gamma_2 \approx 1$ and $\gamma_1 \approx \gamma$. For such catalysts, we can approximate the TOF as

$$r_{tot} = r_1 = k_1 p_{A_2} \left(1 - \gamma\right) \tag{7.15}$$

Because each coverage has an upper limit of 1, the total rate must be limited by both, Equations (7.14) and (7.15) leading to the Sabatier map expressed as

$$r_{tot} = \min\left(r_1, r_2\right) = \min\left(k_1 p_{A_2} \left(1 - \gamma\right), k_2 p_B \left(1 - \sqrt{\gamma}\right)\right) \tag{7.16}$$

Figure 7.6 shows the Sabatier map in comparison to the full solution from the microkinetic model. The Sabatier map gives an excellent description for small approaches to equilibrium ($\gamma \to 0$). There is some discrepancy at the maximum where the Sabatier analysis predicts too high rates. This can be attributed to a failure of describing coverages that are in between 0 and 1, since these are the limiting cases

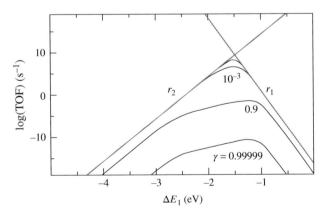

FIGURE 7.6 Sabatier map plotted as a function of ΔE_1 (lines denoted r_1 and r_2 at an approach to equilibrium of 0). The activity maps from the solution to the full microkinetic model at several different approaches to equilibrium are shown for comparison.

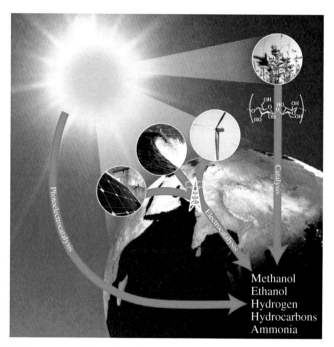

FIGURE 1.1 Illustration of the role of catalysis in providing sustainable routes to fuels and base chemicals. Whether the energy flux from sunlight is harvested through biomass, through intermediate electricity production from photovoltaics or wind turbines, or directly through a photoelectrochemical reaction, the process always requires an efficient catalyst, preferably made of earth-abundant materials. Taken from Nørskov and Bligaard (2013) with permission from Wiley.

Fundamental Concepts in Heterogeneous Catalysis, First Edition. Jens K. Nørskov, Felix Studt, Frank Abild-Pedersen and Thomas Bligaard.
© 2014 John Wiley & Sons, Inc. Published 2014 by John Wiley & Sons, Inc.

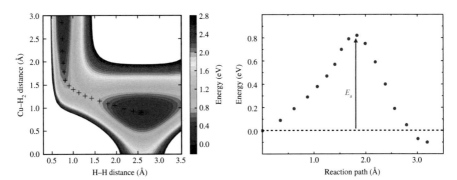

FIGURE 2.2 *Left*: PES for H_2 dissociation over Cu(111). The potential energy of the system is shown as a function of the Cu–H_2 and H–H distance, respectively. H_2 far from the Cu surface has been chosen as a reference. The lowest potential energy path for H_2 splitting is marked with black crosses. *Right*: PED for H_2 dissociation where the lowest potential energy (from the figure on the left) is plotted as a function of the reaction path. The PES is calculated without relaxations of the hydrogen and copper atoms. If these are taken into account, a slightly lower barrier of 0.78 eV is found (see CatApp).

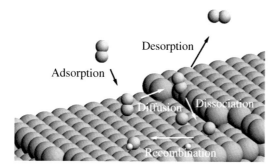

FIGURE 2.4 Illustration of the elementary reaction steps on surfaces.

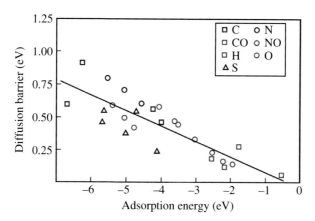

FIGURE 2.8 Diffusion versus adsorption on metal surfaces. The diffusion barriers for a range of different adsorbates are plotted as a function of their adsorption energy. Adapted from Nilekar et al. (2006).

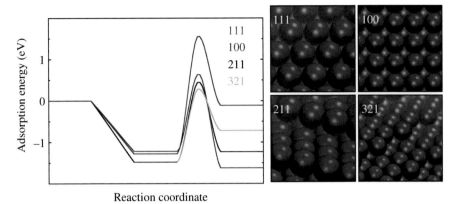

FIGURE 2.15 *Left*: CO dissociation on different facets of fcc Ni. CO is first adsorbed molecularly and then dissociated into adsorbed carbon and oxygen. The barrier for dissociation is extremely structure sensitive being almost an eV higher over the (111) facet than over the more undercoordinated facets. *Right*: illustration of the fcc (111), (100), (211), and (321) facet structures. Adapted from Andersson et al. (2008).

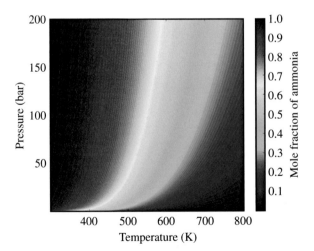

FIGURE 3.1 Equilibrium concentration of ammonia plotted as a function of temperature and pressure assuming a N_2:H_2 ratio of 1:3. Low temperatures and high pressures favor ammonia, whereas high temperatures lead to a shift in equilibrium toward the reactants. Industrially, ammonia synthesis is performed at approximately 700 K and at a total pressure on the order of 100 bar.

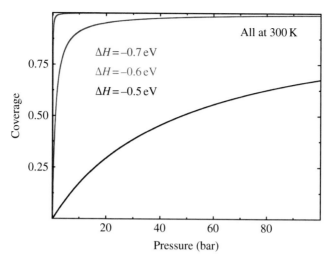

FIGURE 3.5 Coverage of species A plotted as a function of pressure at three different values of adsorption enthalpy, ΔH. The temperature is 300 K and the standard adsorption entropy is -2 meV.

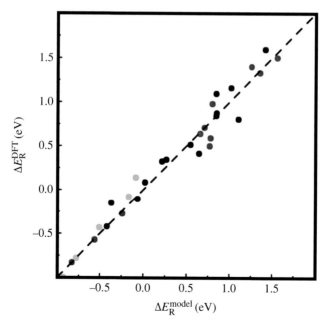

FIGURE 6.4 Graph shows DFT-calculated reaction energies against model reaction estimated using Equation (6.3). The metal-/structure-dependent parameter has been calculated for Pt(111). Reactions involved are dehydrogenation of CH_3OH (black), C_3H_5 (red), CH_3SH (green), cysteine (blue), and ethylene (magenta). Adapted from Abild-Pedersen et al. (2007).

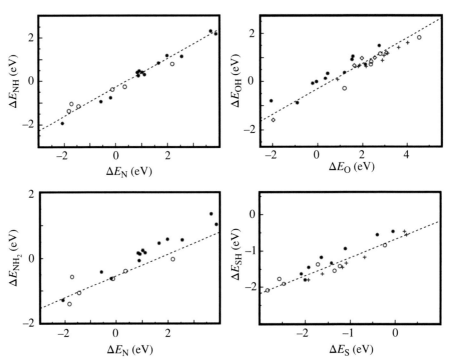

FIGURE 6.6 Figures show NH$_x$ versus N scaling on transition metals and metal nitrides, OH versus O scaling on transition metals and metal oxides, and SH versus S scaling relations on transition metals and metal sulfides. Data points involve stepped transition metal surfaces (black), close-packed transition metal surfaces (blue), and the nitride, oxide, and sulfide surfaces (red). The dashed line shows the best fit to the points using the theoretical slope. Adapted from Fernandez et al. (2008).

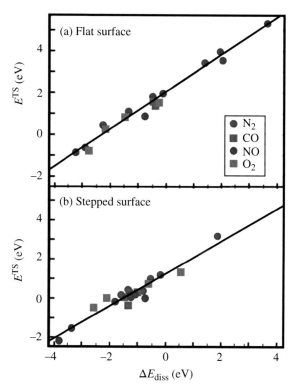

FIGURE 6.9 Linear transition state scaling relations for the dissociation of a number of simple diatomic molecules. It is clear from the plots that for a given surface geometry, all the data cluster around the same "universal" line. Adapted from Nørskov et al. (2002).

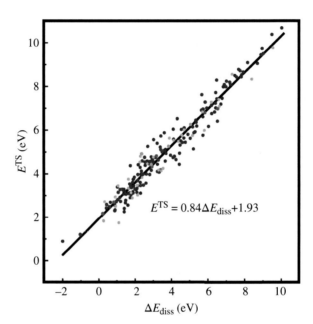

FIGURE 6.10 Universal relation between a large number of coupling reactions on transition metal surfaces. The reactions involved are C–C coupling reactions (blue), C–O coupling reactions (red), C–N coupling reactions (orange), N–N coupling reactions (purple), N–O coupling reactions (green), and O–O coupling reactions (turquoise). Adapted from Wang et al. (2011).

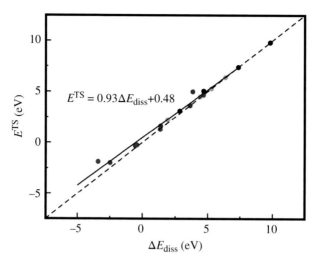

FIGURE 6.11 Universal relation between calculated transition state energies and dissociative chemisorption energies for diatomic systems on (110) transition metal oxide surfaces in the rutile structure. N_2 splitting reactions (black), O_2 splitting reactions (red), CO splitting reactions (orange), and NO splitting reactions (blue) are plotted. Adapted from Vojvodic et al. (2011).

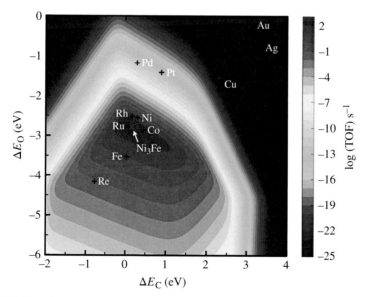

FIGURE 7.9 TOF of CO hydrogenation to methane and water as a function of ΔE_C and ΔE_O. Reaction conditions are 573 K, 40 bar H_2, and 40 bar CO. Values for ΔE_C and ΔE_O for the stepped surfaces of various transition metals are depicted. Taken from Nørskov et al. (2011) with permission from Proceedings of the National Academy of Sciences.

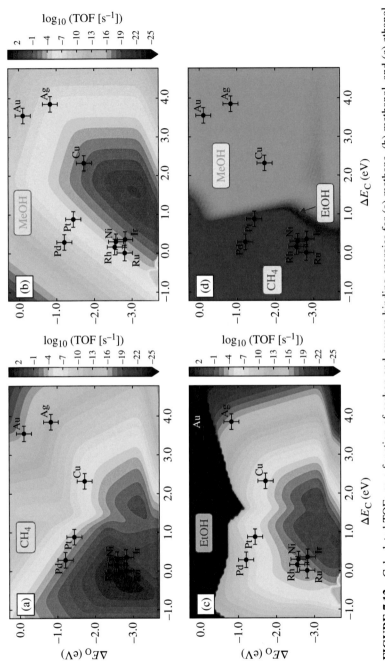

FIGURE 7.12 Calculated TOFs as a function of carbon and oxygen binding energies for (a) methane, (b) methanol, and (c) ethanol formation along with (d) selectivity map. Reaction conditions are at 593 K, 60 bar H_2, and 30 bar CO. The color codes on the selectivity map are determined by weighting the various reaction channels with the selectivity for ethanol, methanol, and methane, respectively. Points represent binding to transition metal (211) surfaces. Taken from Medford et al. (2014) with permission from Springer.

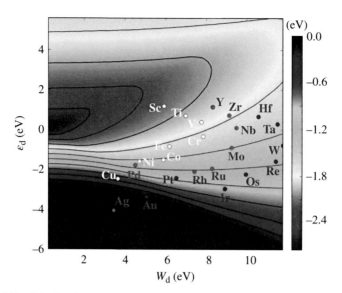

FIGURE 8.7 Calculated change in the sum of the one-electron energies, ΔE_{d-hyb} in eV, using the Newns–Anderson model. The parameters are chosen to illustrate an oxygen $2p$ level ($\varepsilon_a = -5.18$ eV) interacting with the d-states approximated by a semielliptic DOS. The one-electron energies are shown for varying d-band center, ε_d, and varying width of the d-band, W_d. $V_{ad}^2 = 1$ throughout. Most transition metals have been added to the plot using calculated d-band centers and d-band widths. Adapted from Vojvodic et al. (2013).

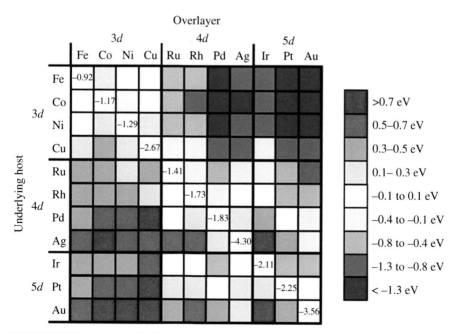

FIGURE 8.15 Calculated shifts in the d-band centers for a number of overlayer structures. The shifts are given relative to the d-band center for the pure metal surface. These shifts therefore reflect the change in reactivity of the overlayer relative to the pure metal. Adapted from Ruban et al. (1997).

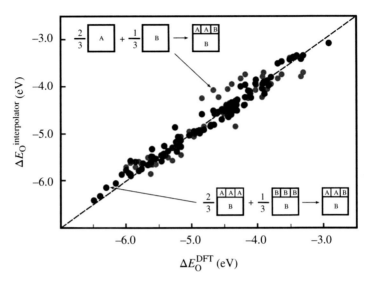

FIGURE 8.17 Oxygen binding energies for a series of surface alloys obtained using two simple interpolation schemes and compared to full DFT calculations. The dashed line indicates perfect agreement. Adapted from Andersson et al. (2006).

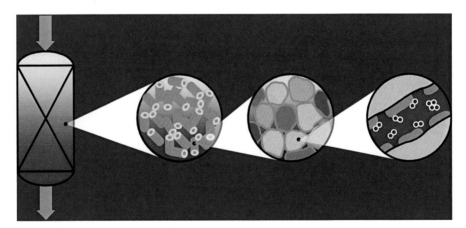

FIGURE 9.1 Illustration of length scales in heterogeneous catalysis from the meter scale of the reactor to the nanometer scale of the catalytic material in a nanometer-sized pore. Taken from Christensen and Nørskov (2008) with permission from The American Institute of Physics.

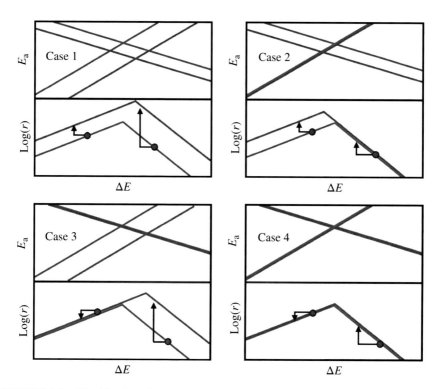

FIGURE 9.5 Classification of the structure dependence of catalytic reactions. In each case, the transition state scaling lines for activation and removal are shown for two different sites. Below the transition state scaling lines, the resulting volcano types are shown. Case 1: both activation and removal exhibit structural dependence. Case 2: activation is independent of structure, but removal shows structural dependence. Case 3: activation is structure dependent but removal is not. Case 4: neither activation nor removal shows structural dependence. Adapted from Nørskov et al. (2008).

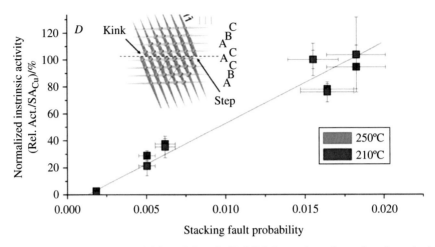

FIGURE 9.10 Catalytic activities of five Cu/ZnO/Al$_2$O$_3$ catalysts in methanol synthesis ($P=60$ bar, $T=210°$, $250\,°C$), normalized to the most active sample as a function of the probability of finding a stacking fault in the Co particles. The insert shows how the stacking fault ends at the surface as a step or a kink. Taken from Behrens et al. (2012) with permission from The American Association for the Advancement of Science.

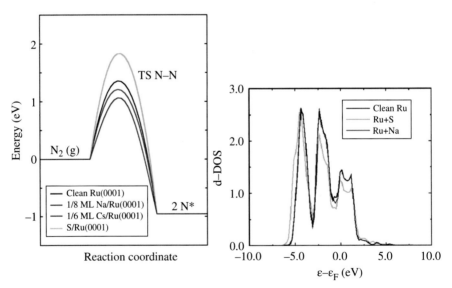

FIGURE 10.3 Potential energy diagram for N$_2$ dissociation on the close-packed Ru(0001) surface in the presence of Na, Cs, and S atoms. To the right, the d-projected DOS for the surface Ru atoms is shown. It is seen that while S atoms shift the d states down, the alkali atoms have essentially no effect. The alkali atoms set up large electrostatic potentials outside the surface, as evidenced by large changes in the work function. Adapted from Mortensen et al. (1998a).

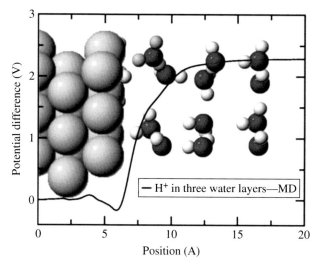

FIGURE 11.3 The electrostatic potential (as seen by an electron) outside a charged Pt(111) slab with three water bilayers outside and one solvated hydronium ion (yellow) per unit cell corresponding to a potential of ca. −2 V versus reversible hydrogen electrode (RHE). The electrostatic potential due to the charged interface is averaged parallel to the surface and is calculated from a time average of a density functional theory (DFT) molecular dynamics simulation of a proton solvated in three water layers obtained after equilibration of the system at 300 K. Adapted from Rossmeisl et al. (2008b).

(a)   (b)

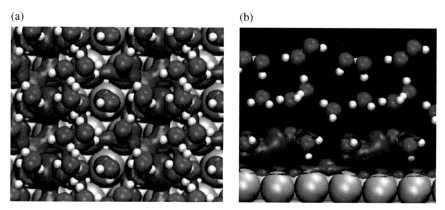

FIGURE 11.4 (a) Top view and (b) side view of a solvated protons in three water layers on top of a Pt(111) electrode. The blue isosurfaces are regions of positive charge around the proton solvated in the water. The purple isosurfaces on the Pt surface are regions of negative charge at the electrode surface. In this case, the proton concentration is very high (one proton per six surface atoms corresponding to a potential of ca. −2 V vs. RHE). Taken from Skulason et al. (2010) with permission from The American Chemical Society.

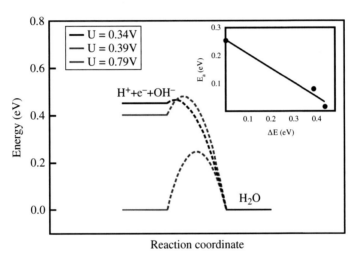

FIGURE 11.8 Potential energy profiles for reducing one OH in the half-dissociated water network by a proton from the water layer at three different potentials. Inset shows the Brønsted–Evans–Polanyi (BEP) relationship for the charge transfer process. The line represents a fit to the data showing a transfer coefficient (γ in Eq. 6.5) of 0.5. Adapted from Tripkovic et al. (2010).

the Sabatier map was constructed for (coverages on the top of activity curves are usually around 0.5, see Fig. 7.2). For larger approaches to equilibrium ($\gamma \rightarrow 1$), however, the optimal catalyst is not defined by the position of the Sabatier map, but moves to more noble surfaces.

The Sabatier analysis can in principle be performed for any reaction. In case that there are more than two reaction steps, the Sabatier volcano could be constructed in analogy to Equation (7.16) by assuming that all intermediates that go in the forward direction have optimal coverages and by calculating the approach to equilibrium for each forward rate from the given approach to equilibrium for the overall reaction (as done in Eq. 7.8 for the reaction earlier) under the assumption that all other partial reactions are in equilibrium. This will give a first approximation to each forward rate of the reaction. The Sabatier volcano provides an upper limit to the total rate by setting this rate equal to the minimum of all forward rates:

$$r_{tot} = \min\left(r_1, r_2, \ldots, r_n\right) \tag{7.17}$$

The Sabatier volcano is in general useful if one wants to obtain an upper limit of the overall rate in cases where there are many competing reaction steps, and it is not a priori clear which steps are rate determining. Its limitation to small approaches to equilibrium, however, should be kept in mind.

7.4 EXAMPLES OF ACTIVITY MAPS FOR IMPORTANT CATALYTIC REACTIONS

7.4.1 Ammonia Synthesis

We will now analyze more complex reactions than the simple generic catalytic reaction discussed earlier, starting with the ammonia synthesis reaction. The free energy diagram of ammonia synthesis has been discussed in Chapter 3 (see Figs. 3.9 and 3.10), and a microkinetic model for this reaction has been developed in Chapter 5. The microkinetic model deals with ammonia synthesis on a stepped Ru surface, and we will now use the same model and extend it to all possible (stepped) metal surfaces through the scaling relations for adsorbates and for transition states as introduced in Chapter 6.

The microkinetic model gives the TOF as

$$R(T,p) = k_1 p_{N_2} \theta_*^2 \left(1 - \gamma\right) \tag{7.18}$$

where k_1 is the forward rate constant for the splitting on N_2 and the approach to equilibrium, γ, is

$$\gamma = \frac{p_{NH_3}^2}{K_{eq} p_{N_2} p_{H_2}^3} \tag{7.19}$$

For the ith reaction step, the equilibrium constant is given by $K_i = e^{-\Delta G_i / k_B T}$. The Gibbs free energy of the reaction step is explicitly dependent on the reaction energy, ΔE_i, through $\Delta G_i = \Delta E_i - T\Delta S_i$, which can be expressed via the scaling relations as

$$\Delta E_i = \sum_{j=1}^{N} \left(\Delta\gamma_j \Delta E^{A_j} \right) + \Delta\xi \tag{7.20}$$

As defined in Chapter 6, $\Delta\gamma_j$ is the change in the scaling parameter for the reaction step considered, ΔE^{A_j} is the variation in binding energies of the base elements, and $\Delta\xi$ is the part that depends on the substrate and thus has to be calculated for a single system.

The reaction energies are thus completely determined by ΔE_N and ΔE_H. For the metals of interest here, the variation in ΔE_H is very modest, so we choose to neglect that. This means that the coverage of free sites is given solely by ΔE_N. The transition state scaling behavior of N_2 splitting is shown in Figure 7.7 for a range of transition metal steps. As for θ_*, k_1 also becomes a function of ΔE_N. Under the assumption that N_2 splitting is the rate-determining step for all ΔE_N, we can express the TOF, R, as

$$R(T, p, \Delta E_N) = \frac{k_B T}{h} e^{\frac{\Delta S^a}{k_B}} e^{\frac{-\Delta E^a}{k_B T}} p_{N_2} \theta_*^2 (1-\gamma) \tag{7.21}$$

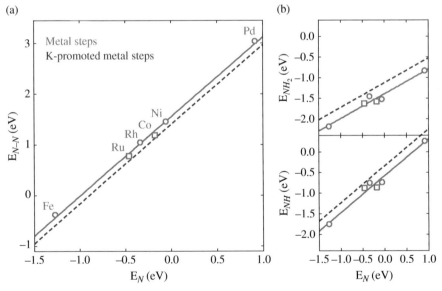

FIGURE 7.7 (a) Transition-state energies of N_2 dissociation plotted as a function of the dissociative chemisorption energy of N_2 over stepped surfaces of various transition metals. (b) Adsorption energy scaling relations for NH and NH_2 intermediates as a function of the nitrogen binding energy. Adapted from *Vojvodic et al.* (2014).

where

$$\theta_* = \cfrac{1}{1+\sqrt{K_2 p_{H_2}} + \cfrac{p_{NH_3}}{\sqrt{K_2 p_{H_2}} K_5 K_6} + \cfrac{p_{NH_3}}{K_2 p_{H_2} K_4 K_5 K_6} + \cfrac{p_{NH_3}}{K_2^{3/2} p_{H_2}^{3/2} K_3 K_4 K_5 K_6} + \cfrac{p_{NH_3}}{K_6}}$$

(7.22)

Figure 7.8 plots the TOF of ammonia synthesis as a function of ΔE_N as obtained from the microkinetic model in conjunction with scaling relations for adsorbates and transition states. A comparison to experimental data is also shown. As can be seen from Figure 7.8, the simple model reproduces the experimental findings qualitatively. Note here that the experimental data is for promoted catalysts. Promotion with, for example, potassium can significantly increase reaction rates by reducing splitting barriers due to electrostatic effects. In the case of ammonia synthesis, potassium also decreases the binding of the adsorbates, which decreases product poisoning and further increases rates. We will discuss effects of promoters in more detail in Chapter 10, but we can see here that the promoted activity map derived from theory resembles the experimental one much more quantitatively.

7.4.2 The Methanation Reaction

Ammonia synthesis is a good example of how one can simplify the description of a reaction network by the use of a single descriptor. Scaling relations always provide a way to reduce the number of independent variables characterizing the catalyst, but in most cases, more than one descriptor is necessary. An important reason is that ΔE_C, ΔE_N, and ΔE_O do not scale well with each other. That means that reactions involving the bonding of, for example, carbon- and oxygen-containing species to the surface will inevitably have two activity descriptors (ΔE_C and ΔE_O).

An example where two descriptors are necessary is the CO methanation reaction:

$$CO + 3H_2 \rightarrow CH_4 + H_2O$$

(7.23)

Another distinct difference from ammonia synthesis is the fact the CO is a strongly adsorbed precursor as compared to N_2, which only binds weakly to most transition metal surfaces. Hence, there is need for including an extra reaction step, as described in Chapter 5. The rate is given by

$$R(p, T, \Delta E_C, \Delta E_O) = k_2 \theta_{CO} \theta_* (1 - \gamma)$$

(7.24)

The full solution of the kinetics using that all reaction barriers and adsorption energy of intermediates scale with ΔE_C and ΔE_O is shown in Figure 7.9.

The activity map for CO hydrogenation shows a single maximum for $(\Delta E_C, \Delta E_O) \approx$ (0.5 eV, −3.0 eV), and it can be seen that Ru and Co the elemental metals are closest to the maximum. This is also what is found experimentally. Ni is used industrially simply because it is cheaper than the other metals.

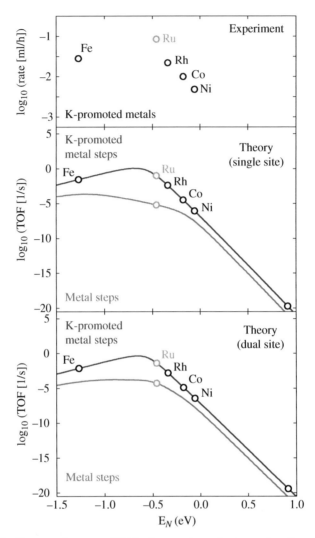

FIGURE 7.8   Turnover frequency (TOF) of ammonia synthesis as a function of the dissociative chemisorption energy of nitrogen. Top panel: Experimental data from *Aika et al.* (1973). Middle panel: Result of the microkinetic model for stepped metal surfaces (blue line). Reaction conditions are 673 K, 100 bar, H_2:N_2 ratio of 3:1, and $\gamma = 0.1$. The effect of potassium promotion has been included (red line). Effects of promotion will be discussed in Chapter 12. Lower panel: Microkinetic model using a two-site model for the adsorption of intermediates. Adapted from *Vojvodic et al.* (2014).

Knowing the optimum value for $(\Delta E_C, \Delta E_O)$ allows a search for other catalysts for this process. In such a computational search, it was predicted that Ni–Fe alloys should be closer to the maximum than Ni and Fe alone. This was confirmed in subsequent experiments (see Fig. 7.10).

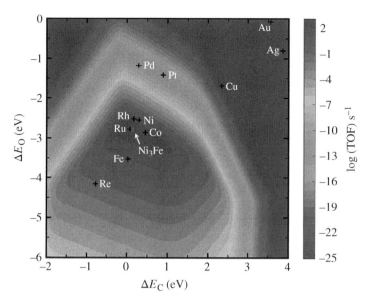

FIGURE 7.9 TOF of CO hydrogenation to methane and water as a function of ΔE_C and ΔE_O. Reaction conditions are 573 K, 40 bar H_2, and 40 bar CO. Values for ΔE_C and ΔE_O for the stepped surfaces of various transition metals are depicted. Taken from Nørskov et al. (2011) with permission from Proceedings of the National Academy of Sciences. (*See insert for color representation of the figure.*)

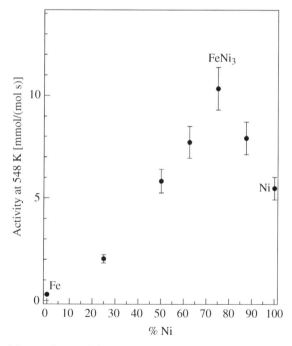

FIGURE 7.10 Measured rate of CO hydrogenation over a series of Fe–Ni catalysts on a spinel support shown as a function of the Ni content. The Ni_3Fe catalyst is clearly better than both pure Ni and pure Fe as suggested by the analysis in Figure 7.9.

TRENDS IN ACTIVITY FOR THIOPHENE HYDRODESULFURIZATION

Hydrodesulfurization (HDS) catalysts are used to remove sulfur and nitrogen from crude oil. The catalysts used are usually based on doped MoS_2, and these systems have been developed and optimized for several decades. In view of stricter legislation regarding sulfur content of transportation fuels, further improvements to these catalysts are needed. Again, it is important to know what the descriptors of catalytic activity are in order to be able to optimize them.

A number of transition metal sulfide catalysts have been studied for the HDS of thiophene (C_4H_4S), which is commonly used as a simple model for the sulfur compounds in crude oil. Typically, there are two types of difficult steps, the breaking of the C–S bond and the removal of adsorbed S from the catalyst surface.

The former is most facile when the S-surface bond is strong, while that makes the latter more difficult. A Sabatier analysis using the adsorption energy of SH as descriptor describes the known trends quite well (see Fig. 7.11). RuS_2 and Co- and Ni-promoted MoS_2 are found to be closest to the top, but there may be room for improvement.

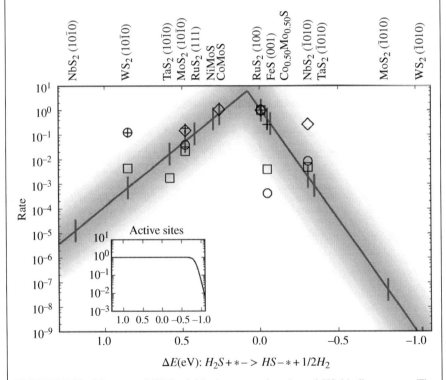

FIGURE 7.11 The rate of HDS of thiophene as a function of HS binding energy. The predicted rate per active site relative to MoS_2 is marked by the solid line, and the vertical lines mark the HS binding energy of different sulfide surfaces. Experimental data which show considerable scatter are also shown. All data are relative to RuS_2. The shaded area indicates the uncertainty in the rates. Taken from Moses et al. (2014) with permission from Springer.

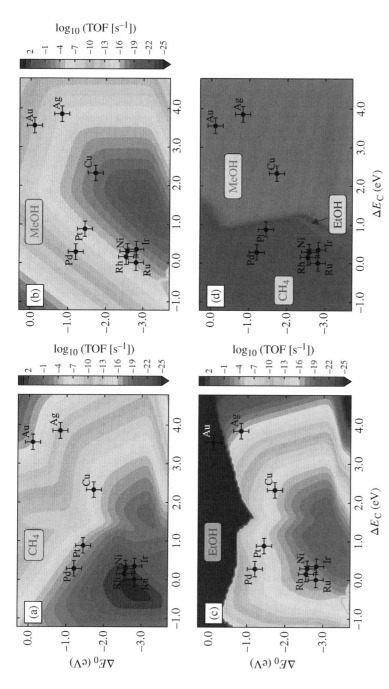

FIGURE 7.12 Calculated TOFs as a function of carbon and oxygen binding energies for (a) methane, (b) methanol, and (c) ethanol formation along with (d) selectivity map. Reaction conditions are at 593 K, 60 bar H_2, and 30 bar CO. The color codes on the selectivity map are determined by weighting the various reaction channels with the selectivity for ethanol, methanol, and methane, respectively. Points represent binding to transition metal (211) surfaces. Taken from Medford et al. (2014) with permission from Springer. (*See insert for color representation of the figure.*)

7.5   SELECTIVITY MAPS

So far, we only discussed the activity of surfaces toward catalyzing certain reactions. Often, there is more than one reaction possible so that a variety of products could be formed. Usually, a high selectivity toward the desired product is wanted, and we will show here how *selectivity maps* can be generated from activity maps where different competing reactions are taken into account.

As an example, we will take the reaction of CO with H_2. As shown earlier, one can generate an activity map for this reaction toward methane as the product, where the TOF is plotted as a function of the carbon and oxygen binding energy. Conversion of CO and H_2 can, however, yield many other products, methanol and ethanol being two of them. In Figure 7.12a–c, we show activity maps for the conversion of CO and H_2 to methane, methanol, and ethanol.

Note that the activity map for methane formation has two distinct maxima, whereas the one in Figure 7.9 only has one. The reason is that in the kinetics leading to Figure 7.9, only the direct formation of methane by CO dissociation and subsequent hydrogenation of adsorbed C and O was included in the analysis. When the synthesis of methanol and ethanol is included in the analysis, new pathways for forming methane appear. The second, lower maximum comes from the pathway where methanol is first formed and the CO bond is broken subsequently. The original maximum found in Figure 7.9 is the dominant one.

From the activity maps, one can form a selectivity map, by calculating the selectivity toward product i *as*

$$S_i = \frac{R_i}{\sum_j R_j}$$

The selectivity map shown in Figure 7.12d shows for which descriptor values the different products dominate. Clearly, there are few values where ethanol is the preferred product, explaining why to date no highly selective higher alcohol synthesis catalyst has been found.

REFERENCES

Aika K, Yamaguchi J, Ozaki A. Ammonia synthesis over rhodium, iridium and platinum promoted by potassium. Chem Lett 1973; 2:161–164.

Medford AJ, Lausche AC, Abild-Pedersen F, Temel B, Schjødt NC, Nørskov JK, Studt F. Activity and selectivity trends in synthesis gas conversion to higher alcohols. Top Catal 2014;57:135–142.

Moses PG, Grabow LC, Fernandez EM, Hinnemann B, Topsøe H, Knudsen KG, Nørskov JK. Trends in hydrodesulfurization catalysis based on realistic surface models. Catal. Lett. 2014; DOI 10.1007/s10562-014-1279-4, to be published.

Nørskov JK, Bligaard T, Abild-Pedersen F, Studt F. Density functional theory in surface chemistry and catalysis. Proc Natl Acad USA 2011;108:937.

Vojvodic A, Medford AJ, Studt F, Abild-Pedersen F, Kahn TS, Bligaard T, Nørskov JK. Exploring the limits: a low-pressure, low-temperature Haber–Bosch process. Chem Phys Lett 2014;598:108–112.

FURTHER READING

Bligaard T, Nørskov JK, Dahl S, Matthiesen J, Christensen CH, Sehested J. The Brønsted-Evans-Polanyi relation and the volcano curve in heterogeneous catalysis. J Catal 2004; 224:206.

Linic S, Jankowiak J, Barteau MA. Selectivity driven design of bimetallic ethylene epoxidation catalysts from first principles. J Catal 2004;224:489.

Loffreda D, Delbecq F, Vigne F, Sautet P. Fast prediction of selectivity in heterogeneous catalysis from extended Brønsted-Evans-Polanyi relations: A theoretical insight. Angew Chem Int Ed 2009;48:8978–8980.

Nørskov JK, Bligaard T, Abild-Pedersen F, Studt F. Density functional theory in surface chemistry and catalysis. Proc Natl Acad Sci USA 2011;108:937.

Raybaud P, Hafner J, Kresse G, Kasztelan S, Toulhoat H. Structure, energetics, and electronic properties of the surface of a promoted MoS_2 catalyst: An ab initio local density functional study. J Catal 2000;190:128.

Sehested J, Larsen KE, Kustov AL, Frey AM, Johannessen T, Bligaard T, Andersson MP, Nørskov JK, Christensen CH. Discovery of technical methanation catalysts based on computational screening. Top Catal 2007;45:9.

8

THE ELECTRONIC FACTOR IN HETEROGENEOUS CATALYSIS

In the previous chapters, we have discussed the potential energy surface on which a surface chemical reaction takes place from several different points of view. We have seen that adsorption energies and activation energies for a given reaction depend on the surface, how there are correlations between these energies, and how these correlations determine trends in reactivity. In this chapter, we will discuss the origin of the surface specificity of adsorption energies and activation energies for surface reactions. We will begin the discussion of how the electronic structure of a surface determines the reactivity, and we will identify the most important electronic structure parameters. We will focus mainly on transition metal catalysts. Many of the concepts are generalizable to other types of catalysts, though, and we will return to that toward the end. This chapter provides a qualitative model of electronic factors in heterogeneous catalysis. A more quantitative mathematical treatment will be presented in Chapter 12.

8.1 THE d-BAND MODEL OF CHEMICAL BONDING AT TRANSITION METAL SURFACES

We will be focusing here on understanding variations in bond energies and activation energies from one transition metal to the next, and we will start by studying variations in the adsorption energy of atomic oxygen. The oxygen atom, $[He]2s^2 2p^4$, has 4 valence electrons in the $2p$ orbitals. As oxygen approaches the surface, these adsorbate

Fundamental Concepts in Heterogeneous Catalysis, First Edition. Jens K. Nørskov,
Felix Studt, Frank Abild-Pedersen and Thomas Bligaard.
© 2014 John Wiley & Sons, Inc. Published 2014 by John Wiley & Sons, Inc.

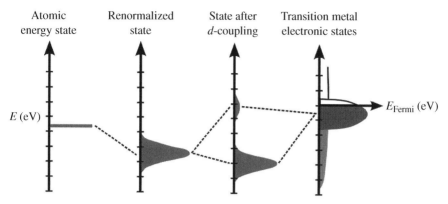

FIGURE 8.1   Schematic illustration of the interaction between an adsorbate valence level and the delocalized *s*-states and localized *d*-states of a transition metal surface.

electronic states will begin to interact with the electronic states of the surface. It is useful to divide the electronic states of transition metal surfaces into two types: the *sp*-bands and the *d*-bands (see Fig. 8.1). The *sp*-bands originate from the metal valence *s* and *p* atomic orbitals that interact to form broad overlapping bands. The valence *d* orbitals of transition metals are more localized than the *s* and *p* valence orbitals, and they therefore interact more weakly and form narrower bands close to the highest occupied state, the Fermi level.

When the oxygen atom approaches the surface, we can consider the coupling to the surface electronic states in two steps. First, consider the coupling to the surface *sp*-states. Bonding states are formed between the metal *sp*-states and the O 2*p* states close to the bottom of the *sp*-bands. Since the states are well below the Fermi level, they are occupied—the O 2*p* states have been filled to form adsorbed O^{2-} (in reality, it is more complex since the filled states have considerable metallic character, but that is not important for this discussion). All transition (and simple) metals have broad *sp*-bands and will have similar bonding characteristics for the *sp* coupling. The charge transfer to the bonding O 2*p* states is associated with a considerable energy gain.

We now include the coupling between the renormalized (by bonding to the metal *sp*-states) O 2*p* and the metal *d*-states. As mentioned earlier, the surface *d*-states are much narrower than the *sp*-states, and the change in electronic structure due to the coupling to the O 2*p* states is similar to the one found between discrete states in a molecular system: bonding and antibonding states are formed below the renormalized O 2*p* states and above the metal *d*-states (see Fig. 8.1). This will give rise to further contributions to the bond energy, and as in molecular systems, the strength of the interaction will depend on the filling of the antibonding states (the bonding states will always be occupied). There is a subtle difference to normal molecular bonds, though. In a molecule, the occupancy of the antibonding states is given by the number of electrons in the system. At a metal surface, there are always electrons available at the Fermi level, and the filling is given by the position of the antibonding states relative to this energy level.

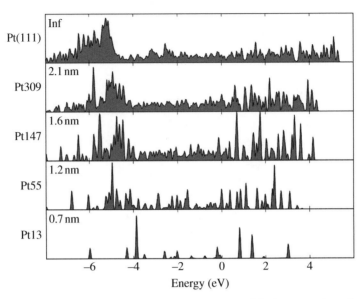

FIGURE 8.2 Calculated *s*-projected density of states (DOS) for a number of platinum clusters in the cuboctahedron structure. The figure clearly shows the transition between a continuous band structure of highly delocalized *s* electrons for large clusters (>2 nm) and the more discrete energy levels for smaller clusters. For comparison, the calculated s-projected DOS for Pt(111) is shown as well. We note the resemblance between the DOS for the Pt$_{309}$ and the Pt(111) slab, suggesting that already at sizes above 2 nm, the band structure is close to converge to the metallic state.

We note that outside a metal surface all adsorbate states that overlap in energy with the metal bands are broadened into resonances. This is a manifestation of the fact that electrons can move between the metal and the adsorbate states; the energy width of the state gives the residence time of the electron on the adsorbate through the Heisenberg uncertainty relations (see also Chapter 11). The broad antibonding states can be partially occupied, and hence, the bond energy can vary continuously as the surface electronic structure changes.

In this picture, we can write the adsorption energy in the simple form:

$$\Delta E = \Delta E_{sp} + \Delta E_{d}. \tag{8.1}$$

Here, ΔE_{sp} is the bond-energy contribution from the free electron-like *sp* electrons, and ΔE_{d} is the contribution from the extra interaction with the transition metal *d* electrons. ΔE_{sp} contributes the most to the bond strength; hence, it is large and negative, whereas ΔE_{d} is a smaller contribution to the bond strength. This is the *d-band model* of adsorption.

In its simplest form, the *d*-band model assumes that ΔE_{sp} is independent of the metal. This is not a rigorous approximation, and it will, for instance, fail when metal particles get small enough that the energy levels do not form a continuous (on the scale of the metal–adsorbate coupling strength) band. Figure 8.2 shows that this, for

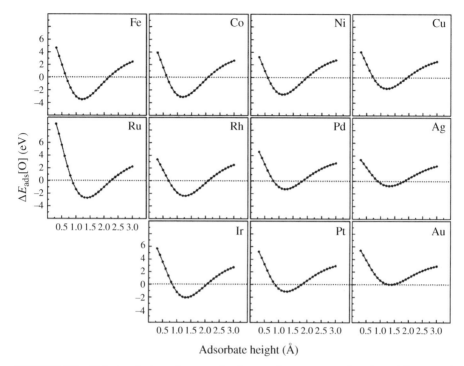

FIGURE 8.3   Calculated adsorption energies for atomic oxygen as a function of distance of the O atom above the surface for a range of close-packed transition metal surfaces (ordered according to their position in the periodic table). The highest coordination surface site was chosen for the adsorption in all cases. All energies are calculated relative to the energy of O_2 in the gas phase, shown as dotted lines in the plots.

instance, happens for very small (below 2 nm) metal particles. In the following, we will work under the assumption that ΔE_{sp} is constant for all transition metals, and we will show that in spite of the approximate nature of the model, it can describe a large number of trends. As mentioned earlier, a more quantitative mathematical treatment is given in Chapter 12.

Consider now the adsorption of atomic oxygen on a range of late transition metal surfaces. We will show how the simple d-band model allows us to understand the trends in variation of the adsorption energy. Figure 8.3 shows calculated adsorption energy profiles as a function of the distance of an O atom above the surface of late transition metals. It can be observed that O binds most strongly on the metals to the left in the transition metal series and stronger to the $3d$ than to the $4d$ and $5d$ metals. This is in excellent agreement with experimental findings, where Ru is observed to bind O much stronger than Pd and Ag, for instance. In addition, Au is observed to be very noble with a binding energy per O atom, which is less than that in the O_2 molecule. The Ag surface is just able to dissociate O_2 exothermically, whereas Cu forms quite strong bonds with O.

Metal *d*-projected DOS

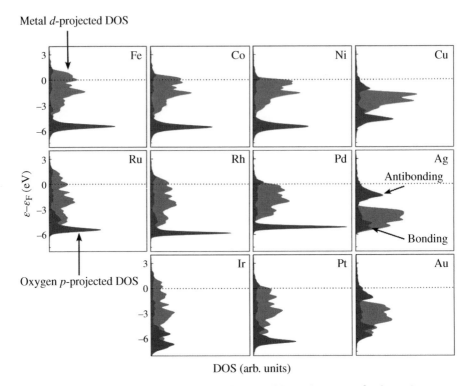

DOS (arb. units)

FIGURE 8.4 The DOS projected onto the *d*-states of the surface atoms for the surfaces considered in Figure 8.3 (light gray). Also shown (dark gray) is the oxygen 2*p*-projected DOS for O adsorbed on the same surfaces. The formation of bonding and antibonding states below and above the metal *d*-states is clearly seen. The O 2*p*-projected DOS shows the weight of a given state on the O 2*p* state. The antibonding states that have more metal *d* character therefore appear weaker than the bonding states that are more O 2*p*-like for all metals except those cases (Ag, Au) where the *d*-states are as low in energy as the O 2*p*-like states. Dashed lines indicate the Fermi level.

Figure 8.4 shows the calculated metal *d*-projected densities of states for the same metal surfaces together with the O 2*p*-projected DOS in the adsorbed state. The formation of bonding and antibonding *p*-states below and above the metal *d*-bands is clearly seen. It can also be observed how the antibonding states for O on Ru are less filled than on Rh, Pd, and Ag, explaining the trend observed in Figure 8.3 that the bonding becomes weaker when you move from left to right in the periodic table.

We have established a picture where the variations in bond strength for O adsorption on transition metals depend on the filling of the antibonding O 2*p*-states. The filling in turn depends on the energy of the antibonding state relative to the Fermi level. In general, this will depend on the distribution of metal *d*-states relative to the Fermi level, the energy of the O 2*p* state (after interaction with the metal *sp*-states, so this is approximately the same for all transition metals), and the strength, V_{ad}, of the O 2*p* and metal *d* coupling.

The simplest possible model includes only the *d*-band center to describe the *d*-band. Figure 8.5 shows a compilation of values of $\varepsilon_d - E_F$ and V_{ad} for all the transition metals. It can be seen that within the 3*d*, 4*d*, and 5*d* series, there are moderate and monotonic variations of V_{ad}. We can therefore expect that the adsorption energy should vary monotonically with $\varepsilon_d - E_F$. This is indeed what is seen for O adsorption on the 4*d* transition metals in Figure 8.6. Both experiments and DFT calculations show that the bonding becomes stronger (i.e., ΔE_{ads} becomes more negative) as we move left in the periodic table and the *d*-states move up in energy relative to the Fermi level. This means that the antibonding states are also moving up in energy and becoming less filled, hence the stronger bonding. The same is observed for the 3*d* and 5*d* series (see Fig. 8.4).

Idealized *d*-band filling

V_{ad}^2(relative to Cu)

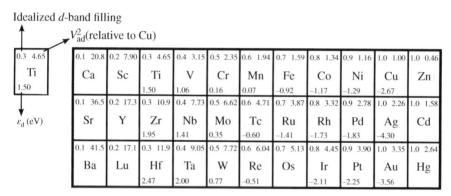

0.3 4.65
Ti
1.50

ε_d (eV)

0.1 20.8	0.2 7.90	0.3 4.65	0.4 3.15	0.5 2.35	0.6 1.94	0.7 1.59	0.8 1.34	0.9 1.16	1.0 1.00	1.0 0.46
Ca	Sc	Ti	V	Cr	Mn	Fe	Co	Ni	Cu	Zn
	1.50	1.06	0.16	0.07	-0.92	-1.17	-1.29	-2.67		
0.1 36.5	0.2 17.3	0.3 10.9	0.4 7.73	0.5 6.62	0.6 4.71	0.7 3.87	0.8 3.32	0.9 2.78	1.0 2.26	1.0 1.58
Sr	Y	Zr	Nb	Mo	Tc	Ru	Rh	Pd	Ag	Cd
		1.95	1.41	0.35	-0.60	-1.41	-1.73	-1.83	-4.30	
0.1 41.5	0.2 17.1	0.3 11.9	0.4 9.05	0.5 7.72	0.6 6.04	0.7 5.13	0.8 4.45	0.9 3.90	1.0 3.35	1.0 2.64
Ba	Lu	Hf	Ta	W	Re	Os	Ir	Pt	Au	Hg
		2.47	2.00	0.77	-0.51		-2.11	-2.25	-3.56	

FIGURE 8.5   Part of the periodic table, showing a number of electronic structure parameters. The *d*-band centers are calculated for the most close-packed surfaces of the experimentally predicted crystal structures fcc(111), hcp(0001), and bcc(110). The idealized *d*-band fillings are shown in the upper left corner, and a coupling matrix element relative to Cu, $V_{ad}^2 / V_{ad/Cu}^2$, between the adsorbate and the metal *d*-states is given as well. The latter is a measure for the extent of the metal *d*-states, and the coupling matrix element to any adsorbate state will be approximately proportional to this number. Adapted from Ruban et al. (1997).

THE *d*-PROJECTED DENSITY OF STATES

The DOS projected onto the *d*-states that interact with the adsorbate state can be characterized by the moments of the d DOS. The first moment is the *d*-band center:

$$\varepsilon_d = \frac{\int_{-\infty}^{\infty} n_d(\varepsilon)\varepsilon \, d\varepsilon}{\int_{-\infty}^{\infty} n_d(\varepsilon) d\varepsilon}$$

and the higher moments ($n > 1$)

$$\varepsilon_d^{(n)} = \frac{\int_{-\infty}^{\infty} n_d(\varepsilon)(\varepsilon - \varepsilon_d)^n \, d\varepsilon}{\int_{-\infty}^{\infty} n_d(\varepsilon) \, d\varepsilon}$$

describe the width and shape in more detail. A simple rectangular model for the d-band takes the form

$$n_d(\varepsilon) = \begin{cases} \dfrac{10}{W_d} & \text{if } \varepsilon_d - \dfrac{W_d}{2} < \varepsilon < \varepsilon_d + \dfrac{W_d}{2} \\[2mm] 0 & \text{elsewhere} \end{cases}$$

where 10 accounts for the total number of d electrons in the band and W_d is the bandwidth. This simple assumption about the form of the d-band brings out the relationship between first and second moment and bandwidth and the fractional filling $f = N_d/10$ of the band such that

$$f = \frac{\int_{\varepsilon_d - W_d/2}^{\varepsilon_F} \dfrac{10}{W_d} \, d\varepsilon}{\int_{\varepsilon_d - W_d/2}^{\varepsilon_d + W_d/2} \dfrac{10}{W_d} \, d\varepsilon} = \frac{1}{2} - \frac{1}{W_d}(\varepsilon_d - \varepsilon_F)$$

If we center the band at ε_d, we can solve for the hybridization part of the bond energy, ΔE_d^{hyb}, such that

$$\Delta E_d^{hyb} = \int_{-\frac{W_d}{2}}^{\varepsilon_F} \frac{10}{W_d} \varepsilon \, d\varepsilon = \frac{5}{W_d}\left(\varepsilon_F^2 - \frac{W_d^2}{4}\right) = 5 W_d f(f-1)$$

The bonding in solids obtained from this simple model is seen to depend on the degree of filling f. In fact, the binding should be maximum for a half-filled d-band. This dependence is exactly what is found experimentally.

The second moment, or the width of the d-band, also affects the interaction energy, but for the late transition metals, this effect is considerably smaller than the effect of varying the d-band center. This is illustrated in Figure 8.7. It shows the result of a model calculation to be described in detail in Chapter 12 of the bond energy for an adsorbate as a function of both the d-band center and the width. In Figure 8.7, we have included values of the d-band widths and centers (W_d, ε_d) for the transition metals.

We note that the reason that the d-band center is a good descriptor of the adsorption energy variations is that it correlates well with the position of the upper d-band edge, which basically controls the position and filling of the antibonding states. If we use a slightly more advanced descriptor, $\varepsilon_d^W = \varepsilon_d + \frac{1}{2}W_d$, the energy of the upper

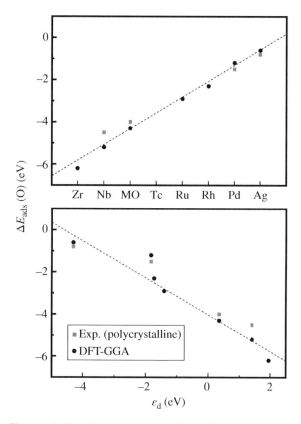

FIGURE 8.6 Changes in the adsorption energy of atomic oxygen along the $4d$ transition metal series. The DFT calculated results are compared to those from experiments. In the graph below, the same data is plotted as a function of the d-band centers taken from Figure 8.5. Adapted from Hammer and Nørskov (2000).

band edge, it provides an even more accurate description of the variations as can be seen in Figure 8.8.

It is clear that the effect of varying the d-band center for a given adsorbate and a fixed coupling matrix element must be similar for any adsorbate with low-lying adsorbate states. This includes H, S, C, N, and many other atomic and molecular adsorbates.

The qualitative picture developed earlier has been verified experimentally. In Figure 8.9 we show X-ray emission and X-ray absorption experiments of N adsorption on Cu(100) and Ni(100) surfaces, combined with DFT calculations of the same systems.

N adsorbs considerably stronger on Ni than on Cu, and this clearly correlates with the filling of the antibonding N $2p$–metal d-states. Since Cu has completely filled d-states (i.e., the d-band is positioned well below the Fermi level), the antibonding states on Cu are mostly filled. On Ni, on the other hand, where the d-bands are pinned to the Fermi level because of the fractional filling of the d-states, the antibonding O $2p$–Ni $3d$ states are partly empty, thus confirming the dependence of the N adsorption-induced changes in the electronic structure on the metal.

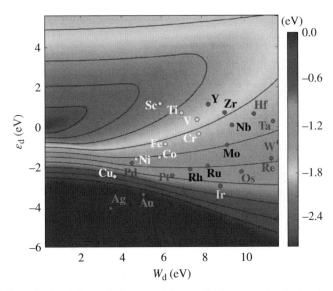

FIGURE 8.7 Calculated change in the sum of the one-electron energies, ΔE_{d-hyb} in eV, using the Newns–Anderson model. The parameters are chosen to illustrate an oxygen $2p$ level ($\varepsilon_a = -5.18\,\text{eV}$) interacting with the d-states approximated by a semielliptic DOS. The one-electron energies are shown for varying d-band center, ε_d, and varying width of the d-band, W_d. $V_{ad}^2 = 1$ throughout. Most transition metals have been added to the plot using calculated d-band centers and d-band widths. Adapted from Vojvodic et al. (2013). (*See insert for color representation of the figure.*)

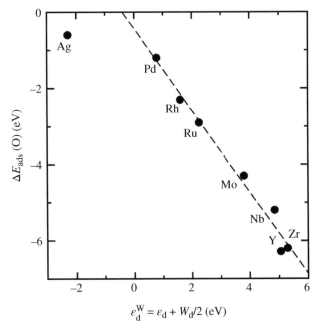

FIGURE 8.8 Calculated oxygen adsorption energies on the most close-packed surface of the $4d$ transition metals plotted against upper band-edge descriptor. The descriptor accounts for the position of the upper edge of the d-band and hence the position of the antibonding state. Adapted from Vojvodic et al. (2013).

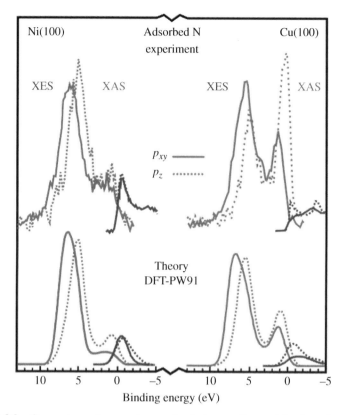

FIGURE 8.9   Comparison of symmetry-resolved nitrogen $2p_{xy}$ and $2p_z$ from (upper) X-ray emission and X-ray adsorption experiments and from (lower) DFT calculations on Cu(100) and Ni(100). Adapted from Nilsson et al. (2005).

We now turn to explain the variations in chemisorption energy between the 3*d*, 4*d*, and 5*d* series. To do this, we will need to include the contribution to the interaction energy from the Pauli repulsion, $\Delta E_d^{\text{ortho}}$, between the adsorbate states and the metal *d*-states. Together with the energy contribution $\Delta E_{d-\text{hyb}}$ from the formation of bonding and antibonding states, this makes up the total energy contribution from the coupling to the metal surface *d* electrons:

$$\Delta E_d = \Delta E_{d-\text{hyb}} + \Delta E_d^{\text{ortho}} \tag{8.2}$$

The Pauli repulsion comes from the energy associated with the orthogonalization of the adsorbate states to the metal states, and this term is proportional to the overlap between the orbitals participating in the bond.

Assuming that the overlap matrix element between the two states scales with the coupling matrix element, we can write the Pauli repulsion in a simple form:

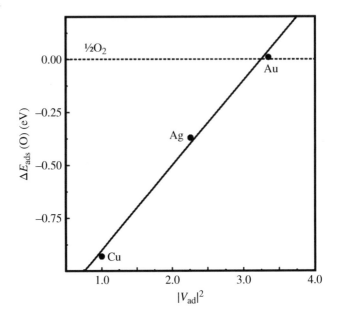

FIGURE 8.10   Variations in the adsorption energies of atomic oxygen (taken from Fig. 8.3) with the size of the coupling matrix element (from Fig. 8.5) for the coinage metals. Adapted from Hammer and Nørskov (2000).

$$\Delta E_d^{ortho} \cong \alpha \, | V_{ad} |^2 \tag{8.3}$$

For oxygen adsorption on Cu, Ag, and Au, we can directly observe ΔE_d^{ortho} because the d-bands are so low-lying that the antibonding states formed between the O atoms and the metal d-states are almost completely filled, meaning that $\Delta E_{d-hyb} \cong 0$. We observe from Figure 8.5 that the coupling matrix element $|V_{ad}|^2$ increases rapidly in going from the $3d$ to the $4d$ and $5d$ transition metals, and in Figure 8.10, we show how the O adsorption energy on the coinage metals Cu, Ag, and Au scales well with $|V_{ad}|^2$.

In this framework, we can understand why Au is so inert with respect to oxygen. For the metals with low-lying, filled d-bands like the coinage metals, the oxygen binding will be weakest. The reason is simple; since both bonding and antibonding adsorbate–metal states are filled, the net effect of the interaction with the metal d-states is entirely repulsive. From Equation (8.3), we see that the repulsive term scales with the strength of the coupling between the adsorbate and the metal surface states; therefore, since Au is the metal with the largest matrix element $|V_{ad}|^2$ of the metals with filled d-bands ($f = 1$ in Fig. 8.5), we find that Au is the metal with the largest Pauli repulsion and the weakest oxygen chemisorption bond, and that is why Ag and Cu are more reactive as seen in Figure 8.10. Considering all of the $f = 1$ metals included in Figure 8.5, Zn should be the pure metal with the strongest oxygen bond, and this is, indeed, found to be the case.

8.2 CHANGING THE *d*-BAND CENTER: LIGAND EFFECTS

For simplicity, let us consider a series of cases where the energy of the renormalized adsorbate state(s), ε_a, and the coupling matrix element between these states and the metal, V_{ad}, can be treated as essentially constant. This is possible if we limit ourselves to situations where a given adsorbate bonds to the same transition metal surface atoms in varying surroundings. In such cases, we would expect that the average energy of the *d* electrons relative to the Fermi level, $\varepsilon_a - \varepsilon_f$, should, to a first approximation, determine the variations in the interaction energy. In the following, we will look at different ways of varying the position of the *d*-band center through local changes in the adsorption site.

Perturbations induced by the atoms surrounding the adsorption site will result in local variations in the *d*-band center. The number of *d* electrons is typically affected less by such changes due to strong Coulomb interactions between the *d* electrons in a given transition metal. In Figure 8.11, we illustrate how a change in *d*-band width of the surface atoms has to lead to a change in the *d*-band center to keep the number of *d* electrons constant.

One way of changing the *d*-band center for a given type of transition metal is by varying the surface structure. As the number of metal neighbors, or the coordination number, of the surface metal atoms changes, then the bandwidth of the *d*-DOS also changes. The rule is that the higher the coordination number, the broader the band. When the coordination number of a surface atom is lowered, the loss in electronic overlap between the atoms in the vicinity creates a local *d*-DOS distribution that is narrower. The consequence of a nearly constant *d*-band filling is an upshift in the *d*-band center, which, according to the *d*-band model, results in a more reactive surface.

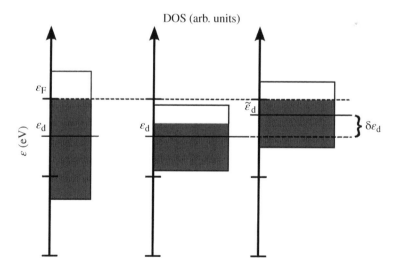

FIGURE 8.11   A schematic illustration of the connection between *d*-band center for a metal with a more than half-filled *d*-shell and the bandwidth for an electronic band with a fixed number of *d* electrons. When the bandwidth becomes narrower, the only way of maintaining the number of *d* electrons fixed is to shift up the center of the band.

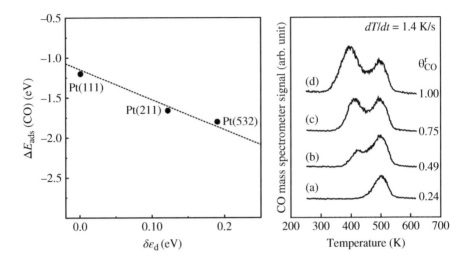

FIGURE 8.12 *Left*: calculated chemisorption energies for CO as a function of the average variation in energy of the *d*-states relative to Pt(111) projected onto the surface atoms to which the adsorbates form bonds. Adapted from Jiang et al. (2009). *Right*: measured thermal desorption spectra for CO on Pt(211). At low coverage, only the strong bonding step sites are occupied, while at high CO coverages, the weaker bonding (111) sites are observed to desorb at ca. 100 K lower temperature. Adapted from Yates (1995).

Take as an example an fcc metal like Pt. In the most close-packed surface structure of Pt, the (111) surface, the surface atoms have a coordination number of 9. For the more open (100) and (110) surfaces, the coordination number decreases to 8 and 7, respectively. Steps and kinks in surfaces and edges and corners on nanoparticles have even lower coordination numbers, from 7 to as low as 5. Figure 8.12 shows how the adsorption energy of CO varies as expected when the *d*-band center changes due to a change in the metal coordination number of the Pt atom to which CO bonds. The close-packed (111) surface binds CO weaker than the step and kink Pt atoms on the (211) and (532) surface by more than 0.5 eV. This has been observed experimentally, for instance, in thermal desorption experiments, where the CO atoms associated with steps desorb at a temperature that is almost 100 K higher than those on the close-packed surface sites (see the right panel in Fig. 8.12).

Another way of changing the band width and hence the center of *d*-states is by straining the surface. Increasing the lattice constant decreases the overlap and decreases the local band width. This can be seen in Figure 8.13 to give a very direct change in bonding to adsorbates, which again scales with the *d*-band center.

Yet another way of changing the *d*-band center in a controlled way is by alloying. The formation of surface alloys can induce changes in the electronic structure, which can be understood in terms of the *d*-band model given that the electronic structure changes are very localized and that the adsorption site remains unchanged.

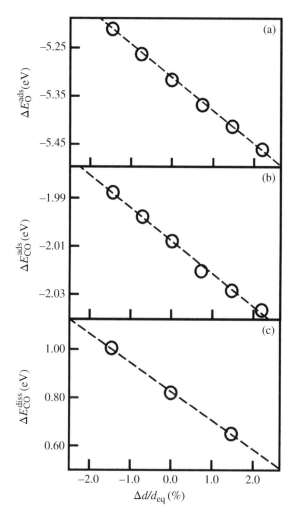

FIGURE 8.13 Effect of strain on the adsorption energy of CO and O and the dissociation barrier for CO on Ru(0001). Adapted from Mavrikakis et al. (1998).

So-called near-surface alloys (NSAs) provide a template system for studying such effects. NSAs or "skins" have been extensively studied as oxygen reduction reaction (ORR) catalysts in PEM fuel cells. If one could reduce the contents of the expensive elemental metal Pt by forming a stable thin layer of the metal on top of a cheap and abundant metal host like Fe, Co, or Ni and improve or maintain its high activity and low overpotential for ORR, it would be a remarkable achievement.

By considering a Pt(111) surface where a series of different $3d$ metals have been sandwiched between the first and second layers, the effect of the subsurface layer of atoms on the reactivity of a Pt(111) overlayer can be studied. The overall effect of the intercalated $3d$ metals is that the d-states of the surface Pt atoms are shifted down in energy as the second layer metal is chosen further to the left in the $3d$ series. In Figure 8.14, O and H adsorption energies have been shown as a function changes in the d-band center induced by the different $3d$ metals sandwiched between the first and second surface layers. The O and H adsorption energies show the same trends: as the d-band center is shifted down in energy, the bond becomes weaker and weaker. For NSAs, the bandwidth changes due to the hybridization between the d-states of the Pt atoms in the surface and the electronic states in the second layer. This indirect interaction between the states intrinsically in the metal can also be termed a ligand effect—the metal ligands of the surface atoms are changed.

Similar effects can be found for metal overlayers. Overlayers of one metal on another are often found for alloy catalysts because one of the components usually segregates to the surface. In such systems, the overlayer atoms have ligand effects from the second layer as in the NSAs, but they also have to adapt to the lattice constant of the host metals. Hence, we have a combination of the ligand and strain effects. Figure 8.15 shows a systematic theoretical study of shifts in d-band centers as different late transition metals are deposited on host materials consisting of other late transition metals.

It can be seen in Figure 8.15 that Pd overlayers on a number of metals except Au and Ag lead to downshifts in the d-band center and hence to weaker adsorption bonds. This has been explored in a set of electrochemical experiments illustrated in Figure 8.16, showing that shifts in the d-band center induced by the underlying metal correlate very well with the observed change in hydrogen adsorption energy.

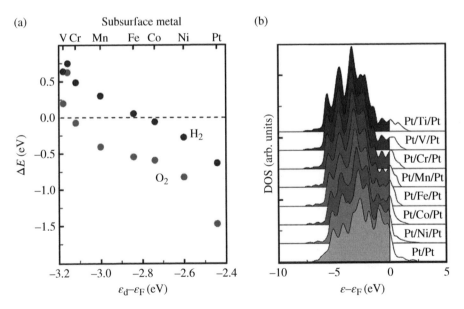

FIGURE 8.14 Calculated changes in the adsorption energy of atomic H and O on a series of Pt-based NSAs. A layer of $3d$ transition metal has replaced the second layer Pt. To the right, the variations in the d-projected DOS for the Pt surface atoms are shown. Adapted from Kitchin et al. (2004).

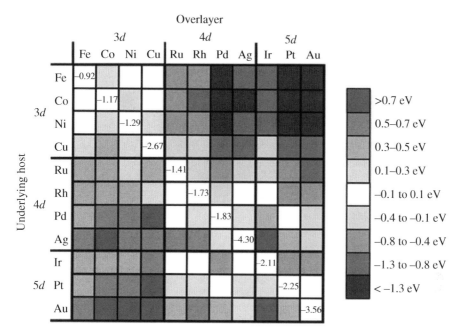

FIGURE 8.15 Calculated shifts in the *d*-band centers for a number of overlayer structures. The shifts are given relative to the *d*-band center for the pure metal surface. These shifts therefore reflect the change in reactivity of the overlayer relative to the pure metal. Adapted from Ruban et al. (1997). (*See insert for color representation of the figure.*)

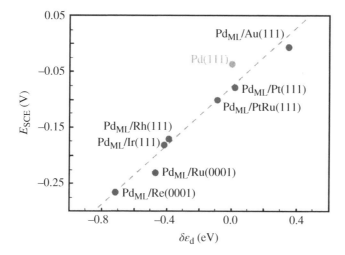

FIGURE 8.16 Variations in the electrochemically measured hydrogen adsorption energies for Pd overlayers on a number of different transition metals as a function of the shift in the *d*-band center calculated from DFT. Adapted from Kibler et al. (2005).

8.3 ENSEMBLE EFFECTS IN ADSORPTION

The cases we have discussed until now have only considered situations where the adsorbate has bonded to one single elemental metal. The variations in the adsorption strength have been induced by second-order effects, meaning that only the surroundings or the atoms in the vicinity of the adsorption site have changed, giving rise to an indirect effect on the adsorbate.

In cases where there are two kinds of metal atoms in the surface, an adsorbate will interact with an ensemble of surface atoms. If an adsorbate interacts with 2 A metal and 1 B metal atom, one would expect an adsorption energy that would be some average of the adsorption energy on metal A and metal B. Such an interpolation principle is found to hold approximately as illustrated in Figure 8.17 showing DFT calculations of the oxygen binding energy for a large number of surface alloys.

The simplest model would be to use data of pure metal surfaces alone and not taking into account strain and ligand effects induced by the host material. Let $A_x B_{1-x}/B$ be the model surface alloy system we want to describe. Here, B is the host material and A is the solute and x is the fractional amount of A in the surface layer. The adsorption energy of oxygen on the surface of our model system can now be approximated as

$$\Delta E\left(A_x B_{1-x}/B\right) = x\Delta E\left(A/A\right) + \left(1-x\right)\Delta E\left(B/B\right) \tag{8.4}$$

This gives a mean absolute error for the data set (light gray points) in Figure 8.17 of about 0.2 eV relative to the full DFT calculations.

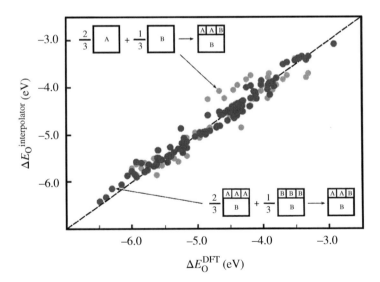

FIGURE 8.17 Oxygen binding energies for a series of surface alloys obtained using two simple interpolation schemes and compared to full DFT calculations. The dashed line indicates perfect agreement. Adapted from Andersson et al. (2006). (*See insert for color representation of the figure.*)

A more accurate interpolation scheme can be obtained from calculations of oxygen adsorption on systems with overlayers of metal A and metal B on the same host material B. This is calculated as

$$\Delta E\left(A_x B_{1-x}/B\right) = x\Delta E(A/B) + (1-x)\Delta E(B/B) \tag{8.5}$$

Using this estimate gives typical mean absolute errors on the data set (dark gray points) in Figure 8.17 of the order 0.1 eV relative to full DFT calculations of the oxygen chemisorption energy.

While the interpolation model is far from perfect, it gives a fast and easy way of estimating the adsorption energies for alloys based on calculations or experiments for simpler systems. Given how simple the two models are, it is surprising how well it works. In fact, the d-band model can be used to elucidate why this is the case.

Under the assumption that ΔE_{sp} in Equation (8.1) is independent of the metal considered, all effects due to having several metal components are to be found in the ΔE_{d-hyb} term. For a system where an adsorbate couples to an ensemble of different surface metal atoms, the natural approach is to assume that the adsorption strength is a linear combination of contributions from each metal. We have seen earlier that ΔE_{d-hyb} is a function of the d-band center, and in a case where the adsorbate couples to several different metal atoms, then according to the arguments earlier, the d-band center of relevance should be an average of the d-band centers for each of the transition metal atoms to which the adsorbate couples:

$$\varepsilon_d \approx \frac{1}{V^2}\sum_j |V_{aj}|^2\, \varepsilon_j; \quad V^2 = \sum_j |V_{aj}|^2 \tag{8.6}$$

Here, V_{aj} is the coupling matrix element between the adsorbate state and the d-states on surface atom j. To the extent that ΔE_{d-hyb} is a linear function of ε_d, variations in the adsorption energy with varying types of surface transition metal atoms will be an average of the contributions from each type.

8.4 TRENDS IN ACTIVATION ENERGIES

So far, we have only focused on electronic structure effects on the chemisorption of intermediates on metal and metal alloy surfaces. To be able to describe the behavior of complete catalytic reaction, we also need information about the activation energies, the energy needed to jump between two intermediate steps on a potential energy surface. The importance of this has been described in some detail in Chapters 6 and 7.

As discussed in the previous chapters, the activation energy, E_a, is defined as the energy of the transition state, ΔE_{TS}, for a given reaction relative to the initial state, ΔE_{IS}:

$$E_a = \Delta E_{TS} - \Delta E_{IS} \tag{8.7}$$

Since the transition state energy is just the "chemisorption energy" of an activated molecule, the arguments behind the d-band should also apply to the interactions in the transition state. It is therefore safe to assume that correlations between the d-band center and transition state energies exist as for chemisorption energies.

In Figure 8.18, we show how the activation energy for methane dissociation on a number of Ni-based surfaces can be correlated with the weighted d-band center as defined in Equation (8.6). The weighted d-band center is necessary since the interaction between the surface and the activated methane complex is seen to have a significant Au component for the NiAu(111) surface alloy structure for the remaining structures the weight from the surrounding surface atoms is negligible and hardly shifts the points. In this approach, it is seen that both alloying (NiAu) and structural effects (compare Ni(111) with Ni(211)) and the effect of strain are described by the weighted d-band center variations.

In fact, the alloying effects can be observed directly in molecular beam scattering experiments monitoring the methane sticking probability as a function of the Au coverage on a Ni(111) surface (see Fig. 8.19).

Structural effects can also be observed experimentally. Again, we illustrate this for the methane activation over Ni surfaces. As expected from the d-band model, the Ni(111) surface has a significantly higher barrier for methane dissociation compared to the stepped Ni(211) surface (see Fig. 8.18). Because sulfur preferentially adsorbs at the step edges on Ni surfaces and since sulfur destabilizes the activated methane complex even more than on the Ni(111) as seen in Figure 8.18, one can experimentally use atomic sulfur to selectively block the more active step sites.

Figure 8.20 shows experimentally measured carbon uptake data during methane activation on a Ni(14 13 13) surface that contains approximately 4% step atoms.

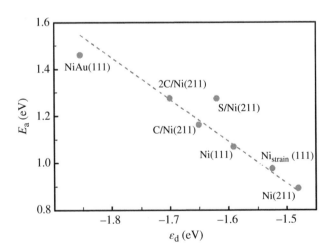

FIGURE 8.18 Calculated variations in the activation energy for methane dissociation over a number of different Ni-based surfaces. The results are shown as a function of the d-band center weighted against the surface metal atoms specifically coupling to the transition state structure. Adapted from Abild-Pedersen et al. (2005a).

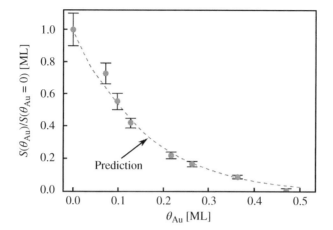

FIGURE 8.19 Measured dissociation probability of methane on Ni(111) surfaces with varying amounts of Au alloying into the surface (sticking is measured relative to the clean surface). The model prediction shown by the dashed line is based on DFT calculations on the change in the activation energy due to the addition of Au atoms into the surface. Adapted from Besenbacher et al. (1998).

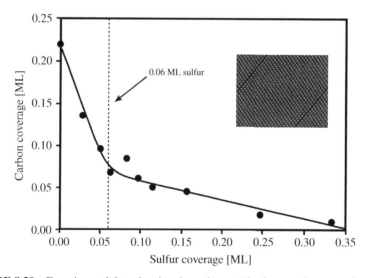

FIGURE 8.20 Experimental data showing the carbon uptake from methane as a function of the sulfur coverage on a Ni(14 13 13) surface. The carbon uptake is a direct measure of the thermal dissociation rate of methane, and it is seen to decrease significantly as S atoms cover the estimated 4% of steps on the Ni(14 13 13) surface (see inset). Adapted from Abild-Pedersen et al. (2005b).

If 2% of S is adsorbed, which corresponds to a half-covered step where all step atoms are blocked by having one S neighbor, the rate of methane dissociation is decreased substantially. This observation shows directly how the undercoordinated step-edge atoms are much faster at dissociating methane than the terrace atoms that are much more densely packed.

Figure 8.18 includes an example where there is an indirect interaction between adsorbates on the surface and the transition state complex. It is observed that preadsorbed carbon at varying coverages affects the dissociation barrier of methane significantly. This is an example where two adsorbates sharing the same surface atom repel each other. Other examples have been shown in Chapter 2 for oxygen adsorption. This effect can be rationalized in the following way: the d-states of the metal atom to which the preadsorbed element couples will be modified by the interaction. In most cases, the interaction broadens the metal d-states, and as a consequence of the constant d filling, the states shift down in energy and the interaction with a second atom or molecule will be weaker.

In Figure 8.18, methane activation on a sulfur-precovered Ni(211) surface is identified as an outlier even when the d-band center is weighted according to Equation (8.6). The d-band model is designed to capture effects from electronic interactions between an adsorbate and the surface alone whether they are direct or indirect. However, there are additional effects due to direct interactions between adsorbates that cannot be described using the d-band model. Sulfur is a large adsorbate with states that have a sizable overlap with the valence orbitals of an incoming molecule. This will give rise to repulsion that is larger than the repulsion originating from the indirect interaction through d-band shifts.

8.5   LIGAND EFFECTS FOR TRANSITION METAL OXIDES

In this chapter, we have concentrated only on the reactivity of transition metal surfaces and shown that trends in adsorption energies and activation energies from one system to the next can be understood in terms of variations in the local d DOS, in particular the d-band center. We end the chapter by indicating that similar concepts can be developed to understand trends in reactivity for transition metal compounds. The point to realize is that if there is a high concentration of electronic states with energies within a few eV (the order of magnitude of adsorbate–surface coupling matrix elements) of the Fermi level, they are likely to dominate variations in reactivity because they can interact with the adsorbate states and form occupied bonding and unoccupied antibonding states.

Consider oxygen adsorption on transition metal oxide surfaces. The role of the narrow bands of metal d-states in the transition metals is now played by localized surface resonances and surface states, typically in the gap of semiconducting or insulating transition metal oxides.

This is illustrated in Figure 8.21a where we consider TiO_2 doped with different transition metals. Doping leads to new d-derived surface states as shown for Cr-doped TiO_2 in Figure 8.21b, and the adsorption energy of O atoms scales with the energy of these states.

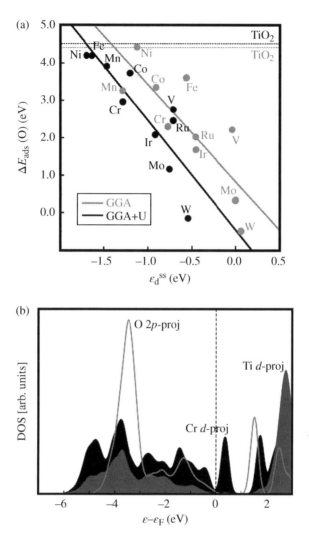

FIGURE 8.21 (a) Correlation between oxygen adsorption energies and the center of the d-projected surface and resonance states on doped TiO_2. The upper solid line is a best linear fit to the DFT calculations, and the lower solid line the best linear fit to DFT + U calculations. (b) Atom-specific projected DOS for Cr-doped TiO_2. Adapted from Garcia-Mota et al. (2013).

REFERENCES

Abild-Pedersen F, Greeley J, Nørskov JK. Understanding the effect of steps, strain, poisons, and alloying: methane activation on Ni surfaces. Catal Lett 2005a;105:9–13.

Abild-Pedersen F, Lytken O, Engbæk J, Nielsen G, Chorkendorff I, Nørskov JK. Methane activation on Ni(111): effects of poisons and step defects. Surf Sci 2005b;590:127–137.

Andersson MP, Bligaard T, Kustov A, Larsen KE, Greeley J, Johannesen T, Christensen CH, Nørskov JK. Towards computational screening in heterogeneous catalysis: paretooptimal methanation catalysts. J Catal 2006;239:501–506.

Besenbacher F, Chorkendorff I, Clausen BS, Hammer B, Molenbroek A, Nørskov JK, Stensgaard I. Design of a surface alloy catalyst for steam reforming. Science 1998;279: 1913–1915.

Garcia-Mota M, Vojvodic A, Abild-Pedersen F, Nørskov JK. Electronic origin of the surface reactivity of transition-metal-doped TiO_2 (110). J Phys Chem C 2013;117:460–465.

Hammer B, Nørskov JK. Theoretical surface science and catalysis—calculations and concepts. Adv Catal 2000;45:71–129.

Jiang T, Mowbray DJ, Dobrin S, Falsig H, Hvolbæk B, Bligaard T, Nørskov JK. Trends in CO oxidation rates for metal nanoparticles and close-packed, stepped, and kinked surfaces. J Phys Chem C 2009;113:10548–10553.

Kibler LA, El-Aziz AM, Hoyer R, Kolb DM. Tuning reaction rates by lateral strain in a palladium monolayer. Angew Chem Int Ed 2005;44:2080–2084.

Kitchin JR, Nørskov JK, Barteau MA, Chen JG. Modification of the surface electronic and chemical properties of Pt(111) by subsurface 3d transition metals. J Chem Phys 2004; 120:10240–10246.

Mavrikakis M, Hammer B, Nørskov JK. Effect of strain on the reactivity of metal surfaces. Phys Rev Lett 1998;81:2819.

Nilsson A, Pettersson LGM, Hammer B, Bligaard T, Christensen CH, Nørskov JK. The electronic structure effect in heterogeneous catalysis. Catal Lett 2005;100:111.

Ruban A, Hammer B, Stoltze P, Skriver HL, Nørskov JK. Surface electronic structure and reactivity of transition and noble metals 1 Communication presented at the First Francqui Colloquium, Brussels, 19–20 February 1996.1. J Mol Catal A: Chem 1997;115:421–429.

Vojvodic A, Nørskov JK, Abild-Pedersen F. Electronic structure effects in transition metal surface chemistry. Top Catal 2013;57:25–32.

Yates JT. Surface chemistry at metallic step defect sites. J Vac Sci Technol A 1995;13:1359.

FURTHER READING

Behm RJ. Spatially resolved chemistry on bimetallic surfaces. Acta Phys Pol A 1998;93: 259–272.

Greeley J, Nørskov JK, Mavrikakis M. Electronic structure and catalysis on metal surfaces. Ann Rev Phys Chem 2002;53:319.

Hammer B, Nørskov JK. Why gold is the noblest of the metals. Nature 1995;376:328.

Kitchin JR, Nørskov JK, Barteau MA, Chen JG. Modification of the surface electronic and chemical properties of Pt(111) by subsurface 3d transition metals. J Chem Phys 2004b; 120:10240–10246.

Liu ZP, Jenkins SJ, King DA. Car exhaust catalysis from first principles: Selective NO reduction under excess O_2 conditions on Ir. J Am Chem Soc 2004;126:10746.

Mehmood F, Kara A, Rahman TS, Henry CR. Comparative study of CO adsorption on flat, stepped, and kinked Au surfaces using density functional theory. Phys Rev B 2009; 79:075422.

Nikolla E, Schwank J, Linic S. Measuring and relating the electronic structures of nonmodel supported catalytic materials to their performance. J Am Chem Soc 2009;131:2747–2754.

Roudgar A, Gross A. Local reactivity of metal overlayers: Density Functional Theory calculations of Pd on Au. Phys Rev B 2003;67:033409.

Vojvodic A, Nørskov JK, Abild-Pedersen F. Electronic structure effects in transition metal surface chemistry. Top Catal 2013;57:25–37.

Xin H, Linic S. Exceptions to the d-band model of chemisorption on metal surfaces: The dominant role of repulsion between adsorbate states and metal d-states. J Chem Phys 2010; 132:221101.

9

CATALYST STRUCTURE: NATURE OF THE ACTIVE SITE

9.1 STRUCTURE OF REAL CATALYSTS

Real heterogeneous catalysts are complex structures typically consisting of several phases. Usually, only one phase is catalytically active, but there are exceptions: for example, one phase may catalyze one part of the total reaction, whereas different phase catalyzes another or the active site is at the boundary between two phases. The active phase usually consists of nanosized particles or is nanostructured in some other way in order to have a large surface area (Fig. 9.1).

Apart from the active phase, many catalysts have a *support* phase, that is, a high surface area material onto which nanoparticles of the active phase are anchored in order to stabilize the high surface area. Typical supports are Al_2O_3, SiO_2, or other materials, which can be prepared in a stable, porous form. In the example in Figure 1.2, the active phase, Ru nanoparticles, is anchored on a boron nitride support.

In other cases, a *structural promoter* is added to keep the nanoparticles of the active phase from sintering together and forming larger particles with a low surface area. In the Fe-based ammonia catalysts used in industry today, Al_2O_3 is added to stabilize the Fe particles.

Finally, there are many cases where *promoters* are added. Promoters are typically materials that spread over the active surface and enhance catalytic activity or selectivity. Taking again ammonia synthesis as the example, the Fe-based catalysts are promoted by K_2O, while the Ru-based catalyst shown in Figure 1.1 is promoted by BaO_x.

Fundamental Concepts in Heterogeneous Catalysis, First Edition. Jens K. Nørskov,
Felix Studt, Frank Abild-Pedersen and Thomas Bligaard.
© 2014 John Wiley & Sons, Inc. Published 2014 by John Wiley & Sons, Inc.

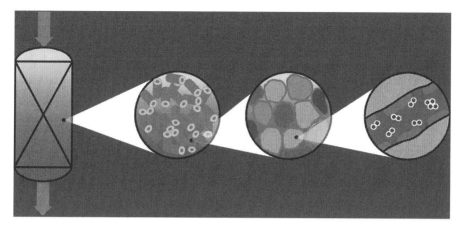

FIGURE 9.1 Illustration of length scales in heterogeneous catalysis from the meter scale of the reactor to the nanometer scale of the catalytic material in a nanometer-sized pore. Taken from Christensen and Nørskov (2008) with permission from The American Institute of Physics. (*See insert for color representation of the figure.*)

Poisons are species on the surface that decrease activity and selectivity. Much care is taken to avoid such species, but that is not always possible. Oxygen is, for instance, a severe poison to the Fe-based ammonia synthesis catalyst, and trace amounts of sulfur in the reactants poison a number of reactions. Sensitivity to poisoning is therefore an important criterion for a good catalyst.

In this chapter, we will discuss a number of aspects of the effect of catalyst structure on catalytic activity, while the topic of poisoning and promotion will be discussed in Chapter 10. First, we will cover the variation in intrinsic catalytic activity of different facets and defects of a surface. Next, we will show how this leads to variations in activity with particle size and shape. This leads us to a close look at the nature of the *active sites* of the surface where most of the catalysis takes place.

9.2 INTRINSIC STRUCTURE DEPENDENCE

It is clear from the discussion of potential energy diagrams in Chapter 2, scaling relations in Chapter 6, electronic structure factors in Chapter 8 that adsorption energies and activation barriers for elementary surface reactions can depend strongly on the local structure of the surface where they take place. In this section, we provide a discussion of the possible consequences of this on the rate for a complete reaction.

There are two ways in which the surface structure of a catalyst can affect the stability of reaction intermediates and the activation energy of an elementary surface chemical reaction. One effect is entirely electronic and the other effect is purely geometrical.

The electronic effect is due to the surface metal atoms in different environments having slightly different local electronic structures. Hence, they interact differently with reactants and intermediates as has been discussed in detail in Chapter 8.

The geometrical effect comes from the fact that different surface geometries provide different configurations to the molecule for bonding. It is in general difficult to differentiate the two effects: steps, for instance, offer atoms with electronic structures different than close-packed surfaces and at the same time offer new surface atom configurations. One way to separate the two effects is by plotting the transition state energy for a surface chemical reaction as a function of the reaction energy for a range of metals and for different surface geometries, as illustrated in Figure 6.8 for CH_4 dissociation and NO dissociation.

For some reactions such as the dissociation of methane (Fig. 6.8a), essentially, the same scaling relation is found for different surface geometries. There can still be electronic step effects—observe, for instance, how the Ni(211) data point is shifted to the left of the Ni(111) data point. This is an example of the d-band effect discussed earlier. The step atoms on the (211) surface (see inset in Fig. 6.8a) have a lower metal coordination number and hence higher lying d-states than the Ni atoms on the close-packed (111) surface. This leads to stronger bonding of the intermediates as well as the transition states. The electronic effect thus corresponds to a displacement along the transition state scaling line. Experimental results illustrating the electronic effect are shown in Figure 9.2.

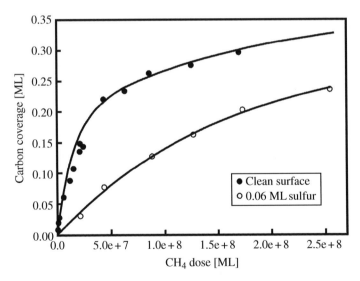

FIGURE 9.2   Measured carbon deposits as a function of CH_4 dose at 500 K on a Ni(14 13 13) surface for two different cases: one case where the surface is "clean" (filled dots) and another where the step sites have been blocked by dosing 0.06 monolayers of sulfur (open circles). The rate of methane dissociation only differs by a factor of ca. 100, which corresponds to a difference in activation barrier of approximately 0.2 eV, in excellent agreement with the results in Figure 6.8a. Adapted from Abild-Pedersen et al. (2005).

For other reactions, there are large shifts in the transition state scaling lines for different geometries. This is shown for NO dissociation in Figure 6.8b. Such strong effects are found generally for N–O, C–O, O–O, N–N, and C–C bond scission. It is found that open surfaces and in particular some kinds of steps have particularly low-lying transition state scaling lines. Experimental results illustrating this effect are shown in Figures 9.3 and 9.4.

We will call these cases, where transition state scaling lines for a certain elementary reaction for different surface geometries differ strongly compared to the electronic effect, as intrinsically structure dependent. The idea is that variations along each of the transition state scaling lines reflect electronic effects as discussed earlier. In this way, we have separated these from the geometrical effects. We can quantify the structure dependence in the following way. If we write the activation energy in terms of the reaction energy in the usual way (Eq. 6.5) as $E_a = \gamma \, \Delta E + \xi$, then strong structure dependence means that the variation $\delta \xi$ in ξ from one structure to the next is stronger than the corresponding variation in the adsorption energy: $\delta \xi > \delta (\Delta E)$. Weak structure dependence is characterized by $\delta \xi < \delta (\Delta E)$.

The geometry dependence of the transition state scaling lines transforms directly into the geometry dependence of the reaction rates, but not in a completely straightforward way. It is not necessarily the case that a lower activation barrier is the same as a higher rate. We will make the analysis for the simple generic reaction used in Chapters 5 and 7 where a single adsorption energy can be used as a descriptor and where there are only two elementary steps to consider. The following discussion is easily generalized to the case with more elementary steps.

Figure 9.5 suggests a classification of structure dependence of catalytic reactions. In the figure, we show transition state scaling lines for activation energies and the

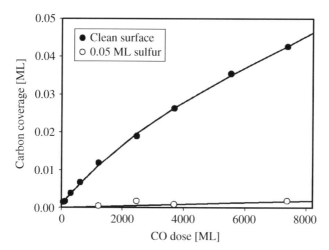

FIGURE 9.3 The experimentally determined carbon coverage on a Ni(14 13 13) single crystal as a function of the CO exposure at 500 K. Results are shown for both the clean surface and for a surface that has been preexposed to 0.05 monolayers of sulfur, which is known to preferentially block the steps. The surface is incapable of dissociating the CO when the steps are blocked. Adapted from Andersson et al. (2008).

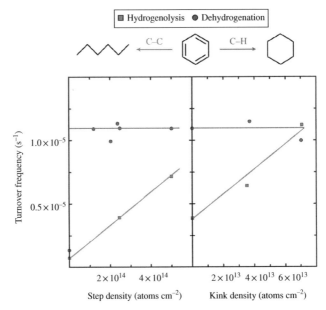

FIGURE 9.4 Measured turnover frequencies of C–C (hydrogenolysis) and C–H (dehydrogenation) rates as a function of step and kink density on Pt surfaces. The much stronger structure dependence of the hydrogenolysis that involves C–C bond formation is clearly seen. Adapted from Blakely and Somorjai (1976).

corresponding Sabatier volcanoes. For two possible rate-determining elementary steps, there are four possibilities depending on whether either elementary reaction shows strong structural effects or not, according to the previous definition. We consider in the following two different local geometries, defining the extremes in the structure dependence of the elementary reactions. The structure with the lowest transition state scaling line (shown red in the figure) could be thought of as belonging to a steplike defect, while the other could represent a close-packed surface.

In Figure 9.5, the arrows indicate what happens to the rate over a given metal when going from a close-packed surface site to a step site where the adsorption energy is more exothermic. The figure suggests a set of rules for determining the nature of the active site. For metals on the right leg of the volcano (noble metals), the steps are always the most active. For the more reactive metals on the left leg, it depends on the degree of structure sensitivity of the elementary step. For Case 3 and Case 4 in Figure 9.5, the site that binds the intermediates most strongly has the lowest rate. This is what one would expect since on the left leg of the volcano, the rate-limiting step is the removal of the adsorbed intermediates. Here, the most reactive sites (in the sense of strongest bonding of surface intermediates) are self-poisoned by the reaction and do not contribute significantly to the catalytic rate. Case 1 and Case 2 in Figure 9.5 are exceptions to this rule. Here, the strongest bonding sites dominate because the reaction barrier is affected more than the reaction energy.

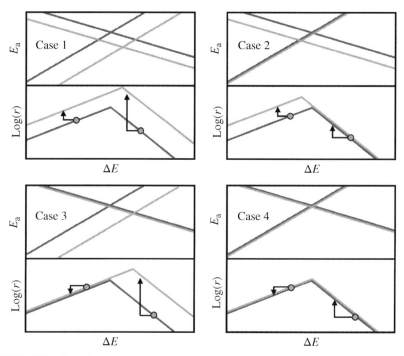

FIGURE 9.5   Classification of the structure dependence of catalytic reactions. In each case, the transition state scaling lines for activation and removal are shown for two different sites. Below the transition state scaling lines, the resulting volcano types are shown. Case 1: both activation and removal exhibit structural dependence. Case 2: activation is independent of structure, but removal shows structural dependence. Case 3: activation is structure dependent but removal is not. Case 4: neither activation nor removal shows structural dependence. Adapted from Nørskov et al. (2008). (*See insert for color representation of the figure.*)

9.3   THE ACTIVE SITE IN HIGH SURFACE AREA CATALYSTS

The structure sensitivity of the reactivity of different facets translates directly into a dependence of the rate and selectivity of supported catalysts on the particle size, morphology, and defect density.

Consider the case where there are two different facets, or possible active sites, m and n, on a nanoparticle catalyst. Assume for simplicity that the rate at each site can be written in an Arrhenius form, $r = v e^{\frac{-E_a}{k_B T}}$. If the two sites have areas A_m and A_n and activation energies E_a^m and E_a^n, and we assume that the prefactor, v, is the same for the two sites, then relative contributions of two sites are given by

$$\frac{r_m}{r_n} = \frac{A_m e^{\frac{-E_a^m}{k_B T}}}{A_n e^{\frac{-E_a^n}{k_B T}}} = e^{-\frac{\left(E_a^m - E_a^n\right) - k_B T\left(\ln(A_m) - \ln(A_n)\right)}{k_B T}} \tag{9.1}$$

This shows that in order to compare the relative importance of different sites, we should compare the geometry probability-weighted activation energies:

$$\tilde{E}_a^i = E_a^i - k_B T \ln\left(A_i\right) \tag{9.2}$$

If one of these is smaller than the rest (on the scale of $k_B T$), this site will dominate the reaction, and the corresponding structure will appear to be *the active site*. Identifying the active site is critical to understanding how to improve a catalyst. It is only if you know the structure and composition of the active site that you can start thinking about how to make it more suited for a reaction. In the following, we will consider examples of elementary reactions where it is possible to identify a well-defined active site.

The particle size dependence is a rough measure of the structure dependence. A convenient measure of the particle size dependence is the relative dependence of the rate per exposed surface area, r, on the particle diameter, d:

$$\alpha = -\frac{d\left(\ln(r)\right)}{d\left(\ln(d)\right)} = -\frac{d}{r}\frac{dr}{dd} \tag{9.3}$$

If all surface sites contribute the same to the rate, then $\alpha = 0$, meaning that the reaction is structure insensitive. A value larger than zero indicates that the active site is of a lower dimensionality than the surface, that is, consists of steps, edges, corners, or kinks. The values of α thus define a *degree of structure sensitivity*.

Using Equations (9.1) and (9.3), we can write the degree of structure sensitivity in terms of the contributions for different surface geometries as

$$\alpha = -\frac{d}{r}\frac{dr}{dd} = -\frac{d}{r}\nabla\sum_i\frac{dA_i}{dd}e^{\frac{-E_a^i}{k_B T}} = -\sum_i\frac{r_i}{r}\frac{d\left(\ln(A_i)\right)}{d\left(\ln(d)\right)} = \sum_i\frac{r_i}{r}\alpha_i \tag{9.4}$$

where α_i is a measure of the dimensionality of the sites of type i ($\alpha_i = 2 - N_{\text{dimensions}}$). Steps and edges (which are one-dimensional) of a particle have $\alpha_i = 1$, whereas corners and kinks (which are point defects and therefore zero-dimensional) have $\alpha_i = 2$. If there were a single, well-defined active site, we would expect an integer value of α. A noninteger value would indicate that several sites contribute. It could also be a consequence of inhomogeneity in the real catalyst. Figure 9.6 shows experimental data for the methanation reaction. Here, CO dissociation is a necessary step in the reaction and as discussed earlier a strongly structure-dependent one. The data in Figure 9.6 indicate that both edges and corners of the metallic nanoparticles contribute to the rate, since the observed α is between one and two.

A number of other reactions belong to the same class of strongly structure-dependent reactions. Ammonia synthesis is a good example. As discussed in Chapters 6 and 7, and

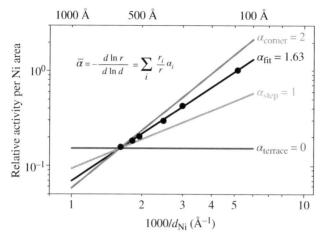

FIGURE 9.6 Double logarithmic plot of the relative CO methanation activity per Ni area of 1% CO in H_2 at 523 K and 1 bar plotted as a function of inverse particle size for a series of Ni catalysts with varying average particle size. The experimental points fit nicely with a line of slope equal to 1.63, indicating structure sensitivity. Adapted from Andersson et al. (2008).

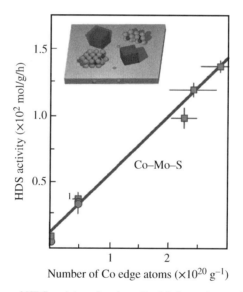

FIGURE 9.7 Scaling of HDS activity of various $Co–MoS_2$ catalysts with the amount of Co atoms at the edge of the S–Mo–S platelets (see inset). Adapted from Topsøe et al. (1996).

expected from discussion of Figure 9.5, ammonia synthesis shows strong structure sensitivity. Another example is shown in Figure 9.7 where hydrodesulfurization (HDS) activity is plotted as a function of the Co edge atoms of a Co–Mo–S catalyst. It can be clearly seen that activity scales with the number of Co edge atoms, which in turn are thus identified as the active site of the Co–Mo–S catalyst.

9.4 SUPPORT AND STRUCTURAL PROMOTER EFFECTS

The support (or structural promoter) can have several effects apart from keeping the nanoparticles of the active component from sintering. First of all, the interaction between the active phase and the support determines the shape of the catalyst particle, as can be seen in Figure 9.8. A weak interaction means that the nanoparticle will have a form that is essentially unperturbed by the support. Such an example is also shown in Figure 1.1. If the interaction is stronger, the interface area will tend to be larger to gain more of the interaction energy, and a truncated particle shape results. An example is shown in Figure 9.9 for palladium nanoparticles that have been grown on an alumina film.

A structural promoter can also affect the surface structure by introducing bulk defects that anchor surface defects such as steps that have a high activity. This is, for instance, the case in the industrial methanol synthesis catalyst. The catalyst is based on Cu nanoparticles, and the steps are found to be the most active for synthesis (Fig. 9.10).

The support determines the particle size by controlling the degree to which the individual particles aggregate to form larger particles. In that way, the support indirectly affects the activity as discussed earlier.

The support can also be a *direct participant* in the reaction. This can happen in three ways:

1. The support may have a catalytic activity of its own, such that it will transform one or more of the products at the other active phase into a new product.

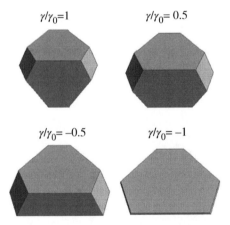

$\gamma/\gamma_0=1$　　　$\gamma/\gamma_0=0.5$

$\gamma/\gamma_0=-0.5$　　　$\gamma/\gamma_0=-1$

FIGURE 9.8 Geometries for a supported nanoparticle for different values of the ratio of γ, the interface free energy minus the surface free energy of the support, and γ_0, the surface energy of the active phase $\left(\dfrac{\gamma}{\gamma_0}\right)$. The value of γ determines the change in energy of the system per interface area, while γ_0 determines the change in energy per surface area of the active phase. $\gamma = \gamma_0$ means that the surface of the particle is unaffected by the presence of the support, while $\gamma = -\gamma_0$ means that the active phase will spread completely over the surface of the support, thus forming a single layer. Adapted from Clausen et al. (1994).

(a)   (b)

(c)

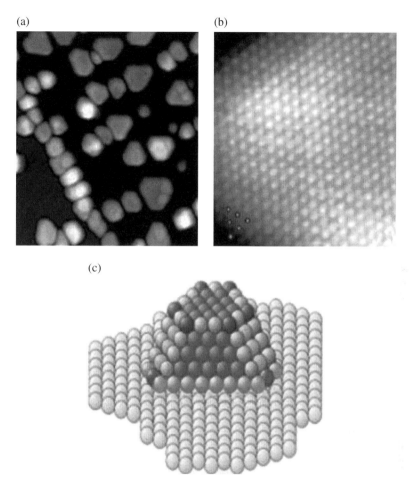

FIGURE 9.9   (a) Image of Pd nanoparticles nucleated at steps and domain boundaries of an alumina film grown on NiAl(000). (b) Atomic resolution images of crystalline Pd nanoparticles. The resolution is kept a few layers down the sides, allowing identification of the side facets. The dots indicate atomic positions consistent with a (111) facet. (c) Schematic representation of a crystalline truncated cuboctahedron of Pd on an oxide surface. The various potential adsorption sites are indicated by coloring in different gray-scales. Adapted from Freund (2008).

This, for instance, happens with methanol synthesis, where some supports with acid–base catalytic activity may transform the methanol produced over a Cu catalyst into dimethyl ether.

2. The support can participate at the interface. The interface sites can have chemical properties that are different from those of both phases. This may be one of the reasons gold particles supported on TiO_2 have catalytic properties different from those of bulk (unsupported) gold.

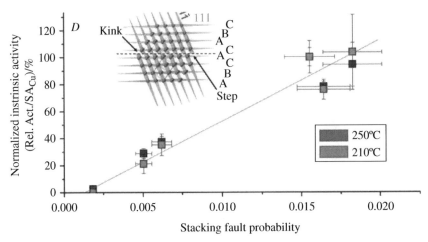

FIGURE 9.10 Catalytic activities of five Cu/ZnO/Al₂O₃ catalysts in methanol synthesis ($p = 60$ bar, $T = 210°C$, $250°C$), normalized to the most active sample as a function of the probability of finding a stacking fault in the Co particles. The inset shows how the stacking fault ends at the surface as a step or a kink. Taken from Behrens et al. (2012) with permission from The American Association for the Advancement of Science. (*See insert for color representation of the figure.*)

3. The support may migrate onto the active phase and alter the properties there. Such an effect, also known as strong metal–support interactions (SMSI), can change the activity in several ways. The support may act as a promoter, as will be discussed later. The support may also block certain sites on the active phase and in that way induce changes in the activity or selectivity. We have, for instance, seen that C–C bond breaking and formation are more affected by the presence of steps than C–H bond breaking and formation. If the support preferentially migrates to the steps, this would change the selectivity between dehydrogenation and hydrogenolysis, for instance.

REFERENCES

Abild-Pedersen F, Lytken O, Engbæk J, Nielsen G, Chorkendorff I, Nørskov JK. Methane activation on Ni(111): effects of poisons and step defects. Surf Sci 2005;590:127.

Andersson MP, Abild-Pedersen F, Remediakis I, Bligaard T, Jones G, Engbæk J, Lytken O, Horch S, Nielsen JH, Sehested J, Rostrup-Nielsen JR, Nørskov JK, Chorkendorff I. Structure sensitivity of the methanation reaction: H₂ induced CO dissociation on nickel surfaces. J Catal 2008;255:6–19.

Behrens M, Studt F, Kaatkin I, Kuhl S, Havecker M, Abild-Pedersen F, Zander S, Girgsdies F, Kurr P, Kniep B, Tovar M, Fischer RW, Nørskov JK, Schlögl R. The active site of methanol synthesis over Cu/ZnO/Al₂O₃ industrial catalysts. Science 2012;336:893–897.

Blakely DW, Somorjai GA. The dehydrogenation and hydrogenolysis of cyclohexane and cyclohexene on stepped (high miller index) platinum surfaces. J Catal 1976;42:181–196.

Christensen A, Nørskov JK. A molecular view of heterogeneous catalysis. J Chem Phys 2008;128:182503.

Clausen BS, Schiøtz J, Gråbæk L, Ovesen CV, Jacobsen KW, Nørskov JK, Topsøe H. Wetting/non-wetting phenomena during catalysis: evidence from in situ on-line EXAFS studies of Cu-based catalysts. Top Catal 1994;1:367–376.

Freund H-J. Model systems in heterogeneous catalysis: selectivity studies at the atomic level. Top Catal 2008;48:137–144.

Hansen KH, Worren T, Stempel S, Lægsgaard E, Baumer M, Freund H-J, Besenbacher F, Stensgaard I. Palladium nanocrystals on Al_2O_3: structure and adhesion energy. Phys Rev Lett 1999;83:4120–4123.

Nørskov JK, Bligaard T, Hvolbæk B, Abild-Pedersen F, Chorkendorff I, Christensen CH. The nature of the active site in heterogeneous metal catalysis. Chem Soc Rev 2008;37:2163.

Topsøe H, Clausen BS, Massoth FE. Hydrotreating catalysis, science and technology. In: Anderson JR, Boudart M, editors. *Catalysis, Science and Technology*. Volume 11, Berlin: Springer; 1996.

FURTHER READING

Boudart M. Catalysis by supported metals. Adv Catal 1969;20:153.

Madix RJ, Friend CM. Gold's enigmatic surface. Nature 2011;479:482.

Nørskov JK, Bligaard T, Hvolbæk B, Abild-Pedersen F, Chorkendorff I, Christensen CH. The nature of the active site in heterogeneous metal catalysis. Chem Soc Rev 2008;37:2163.

Somorjai GA, Joyner RW, Lang B. Reactivity of low index [(111) and (100)] and stepped platinum single-crystal surfaces. Proc R Soc Lond Ser A 1972;331:335.

Yates JT. Surface-chemistry at metallic step defect sites. J Vac Sci Technol A 1995;13:1359.

Zambelli T, Wintterlin J, Trost J, Ertl G. Identification of the "active sites" of a surface-catalyzed reaction. Science 1996;273:1688–1690.

10

POISONING AND PROMOTION OF CATALYSTS

We have already discussed how the support may migrate to the surface of the active phase and affect its reactivity. This is part of a general set of phenomena called promotion and poisoning of the catalyst, depending on whether the selectivity or rate goes up or down or whether the substance is added deliberately or not.

One famous example of a promoter is the effect that alkali metals have on Fe-based ammonia synthesis catalysts. Figures 10.1 and 10.2 show the effect on the rate of dissociation of N_2 and on the total rate. Clearly, part of the effect of adding potassium to the surface is to increase the rate of the rate-limiting step: N_2 dissociation.

The promoting effect of alkali atoms on the dissociation rate for N_2 is also clearly reflected in the potential energy diagram for N_2 dissociation, as depicted in Figure 10.3. Coadsorbed alkali atoms lower the activation barrier for dissociation substantially. The reason is quite simple. The very electropositive alkali atoms donate electrons to the surface, hence becoming partially positive. The alkali-induced increase in positive charge outside the surface and corresponding negative charge in the surface sets up an electrostatic field outside the surface (see Fig. 10.4). This field interacts with the dipole moment of the N_2 molecule in the transition state. Since the molecule is slightly negative (due to transfer of electrons into the antibonding $2\pi^* N_2$ state), the interaction between the induced field and the molecular dipole becomes attractive.

Fundamental Concepts in Heterogeneous Catalysis, First Edition. Jens K. Nørskov,
Felix Studt, Frank Abild-Pedersen and Thomas Bligaard.
© 2014 John Wiley & Sons, Inc. Published 2014 by John Wiley & Sons, Inc.

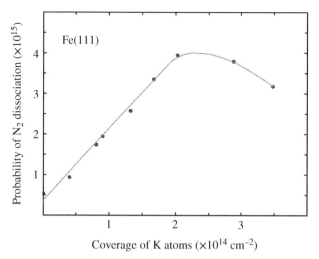

FIGURE 10.1   Measured probability of N_2 dissociation on a Fe(111) surface as a function of the coverage of K atoms on the surface. Adapted from Ertl et al. (1982).

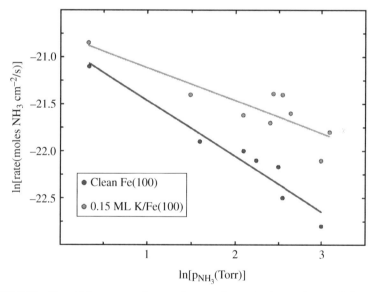

FIGURE 10.2   Plot of the measured rate of ammonia synthesis on clean and K-promoted Fe(100) surfaces as a function of product pressure. Adapted from Strongin and Somorjai (1988).

The interaction energy between an adsorbate and an electrical field ε can be written as

$$\Delta E = \mu\varepsilon - \frac{1}{2}\alpha\varepsilon^2 + \dots \tag{10.1}$$

where μ is the dipole moment of the adsorbate and α is the polarizability.

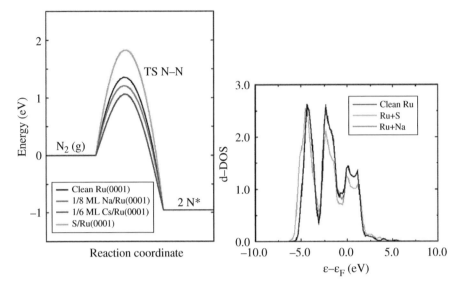

FIGURE 10.3   Potential energy diagram for N_2 dissociation on the close-packed Ru(0001) surface in the presence of Na, Cs, and S atoms. To the right, the d-projected DOS for the surface Ru atoms is shown. It is seen that while S atoms shift the d states down, the alkali atoms have essentially no effect. The alkali atoms set up large electrostatic potentials outside the surface, as evidenced by large changes in the work function. Adapted from Mortensen et al. (1998a). (*See insert for color representation of the figure.*)

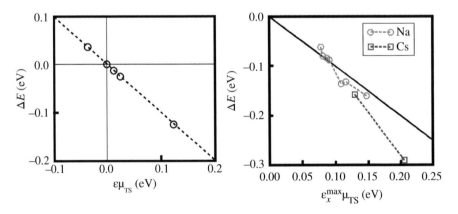

FIGURE 10.4   *Left*: interaction energies between the transition state for N_2 dissociation and a homogeneous electric field, as a function of the electrical field strength ε. The induced dipole moment of the transition state is $\mu_{TS} = -0.11$ eÅ. *Right*: interaction energies between alkali atoms and N_2 in the transition state for dissociation as a function of the maximum field strength induced by the adsorbed alkali atom, ε_x^{max}. The alkali atoms are Na (circles) and Cs (squares). Adapted from Mortensen et al. (1998b).

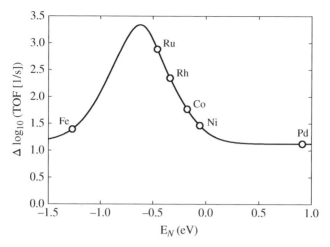

FIGURE 10.5 Difference of the turnover frequency of ammonia synthesis between promoted and unpromoted transition metal catalysts as a function of the nitrogen adsorption energy. The difference is based on the volcanoes shown in Figure 7.8. The promotional effect is most pronounced for metals having an intermediate nitrogen binding energy as two effects take place (decrease of the N-N splitting barrier and destabilization of adsorbed NH*). Note that the plot does not take into account that the number of catalytically active sites does decrease upon promotion as the promoter is blocking some available sites. The experimentally observed difference will hence be somewhat less pronounced. Adapted from Vojvodic et al. (2014). See also Dahl et al. (2001).

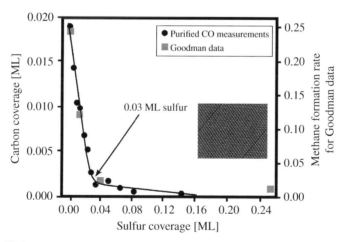

FIGURE 10.6 Experimental data showing how the thermal dissociation rate for CO (as measured by the C uptake (circles) or methane formation (squares)) decreases rapidly as S atoms cover the 4% of steps on a Ni(14 13 13) surface. Adapted from Andersson et al. (2008).

In fact, in the ammonia synthesis reaction, there is an additional promoting effect due to the addition of an alkali (see Fig. 7.7). Intermediate NH_x species formed during the ammonia synthesis reaction are destabilized at the surface. The H atoms tend to be slightly positive, and hence, the molecular dipole moment has the opposite sign. This gives rise to a repulsion between the NH_x and alkali atoms, which lowers the coverage of the latter and hence frees up more sites for N_2 dissociation. Both effects work together to increase the rate, as shown in Figure 10.5. Note that in reality the alkali atoms are bound to O or N atoms on the surface, but the effect is still electrostatic because, if anything, the O and N atoms increase the charge transfer from the alkali atoms.

If, instead of a promoter, you add an atom or molecule to the surface that interacts repulsively with the transition state of a rate-limiting step, thus increasing the activation barrier and decreasing the rate, such an additive is called a poison. In Figure 10.3, it can be seen that S atoms poison the N_2 dissociation process, which occurs mainly because S atoms shift the d states of Ru down in energy.

The degree of poisoning becomes particularly problematic if a process is strongly structure dependent and the poison is attracted to the most active phase. Often, steps are the active sites, and S atoms are strongly attracted to these sites. In that case, a few percent of S atoms on the surface can completely poison a reaction. An example is shown in Figure 10.6.

REFERENCES

Andersson MP, Abild-Pedersen F, Remediakis IN, Bligaard T, Jones G, Engbæk J, Lytken O, Horch S, Nielsen JH, Sehested J, Rostrup-Nielsen JR, Nørskov JK, Chorkendorff I. Structure sensitivity of the methanation reaction: H_2 induced CO dissociation on nickel surfaces. J Catal 2008;255:6–19.

Dahl S, Logadottir A, Jacobsen CJH, Nørskov JK. Electronic factors in catalysis: the volcano curve and the effect of promotion in catalytic am-monia synthesis. Appl Catal A 2001;222:19–29.

Ertl G, Lee SB, Weiss M. Adsorption of nitrogen on potassium promoted Fe(111) and (100) surfaces. Surf Sci 1982;114:527–545.

Mortensen JJ, Hammer B, Nørskov JK. Alkali promotion of N_2 dissociation over Ru(0001). Phys Rev Lett 1998a;80:4333–4336.

Mortensen JJ, Hammer B, Nørskov JK. A theoretical study of adsorbate-adsorbate interactions on Ru(0001). Surf Sci 1998b;414:315–329.

Strongin DR, Somorjai GA. The effects of potassium on ammonia synthesis over iron single-crystal surfaces. J Catal 1988;109:51–60.

Vojvodic A, Medford AJ, Studt F, Abild-Pedersen F, Khan TS, Bligaard T, Nørskov JK. Exploring the limits: a low-pressure, low-temperature Haber–Bosch process. Chem Phys Lett 2014;598:108–112.

FURTHER READING

Koel BE, Kim J. Promoters and poisons. In: Ertl G, Knözinger H, Schüth F, Weitkamp J, editors. Handbook of Heterogeneous Catalysis. Weinheim: Wiley-VCH Verlag GmbH; 2008. p 1593.

Nørskov JK, Holloway S, Lang ND. Microscopic model for the poisoning and promotion of adsorption rates by electronegative and electropositive atoms. Surf Sci 1984;137:65.

Zhang CJ, Hu P, Lee MH. A density functional theory study on the interaction between chemisorbed CO and S on Rh(111). Surf Sci 1999;432:305.

11

SURFACE ELECTROCATALYSIS

In this chapter, we will consider electrocatalytic processes. It is not the intention to provide a complete description of surface electrocatalysis but rather to show how catalytic processes at electrode surfaces can be understood in much the same terms as other surface-catalyzed chemical processes. In particular, we will show that free energy diagrams, scaling relations, and activity maps are tools that are just as useful to analyze trends in electrocatalytic processes as for other heterogeneous catalytic processes.

We note that electrocatalysis and photocatalysis are often closely related phenomena. Figure 11.1 illustrates the working principles of a photocatalytic device for water splitting based on a semiconductor, which can absorb sunlight, and two electrocatalysts (one or both of which could in principle be the same material as the semiconductor). One of the catalysts uses the excited electrons in the conduction band to form molecular hydrogen:

$$2(e^- + H^+) \rightarrow H_2 \tag{R11.1}$$

while the other enables the holes in the valence band to split water to form molecular oxygen and protons:

$$4h^+ + 2H_2O \rightarrow O_2 + 4H^+ \tag{R11.2a}$$

or, equivalently,

Fundamental Concepts in Heterogeneous Catalysis, First Edition. Jens K. Nørskov, Felix Studt, Frank Abild-Pedersen and Thomas Bligaard.

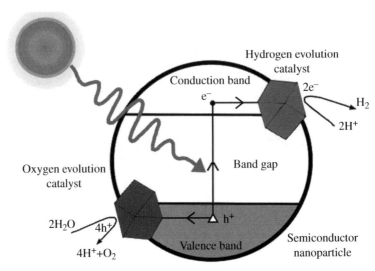

FIGURE 11.1 Illustration of the function of a semiconductor-based solar water splitting device.

$$2H_2O \rightarrow O_2 + 4(e^- + H^+) \tag{R11.2b}$$

Finding good electrocatalysts is therefore one of the core challenges in solar fuel production.

Before getting to the description of trends in electrocatalytic activity, we will discuss the features of surface electrocatalysis that are different from ordinary gas-phase heterogeneous catalysis.

11.1 THE ELECTRIFIED SOLID–ELECTROLYTE INTERFACE

In the following, we will concentrate on water as the electrolyte, but most of what we will be discussing can be generalized to other electrolytes. We will also focus on processes involving proton transfer. The reason is that most of the processes of interest in energy transformations are associated with proton transfers to or from the surface. Again, the majority of what will be discussed is easily transferable to other types of processes.

Figure 11.2 shows a schematic illustration of an electrochemical cell. The potential difference between the anode and the cathode gives rise to a variation in the electrostatic potential through the cell. Since the electrolyte is conducting, there is no electrical field there and the potential is constant. The potential variation happens in the so-called dipole layer close to the two electrodes and sets up strong electrical fields there. The field is set up by the electrons in the electrode (or holes for a positive electrode) and the counterions in the electrolyte—the total charge in the surface and in the screening layer in the electrolyte is the same (otherwise, there would be a field in the electrolyte).

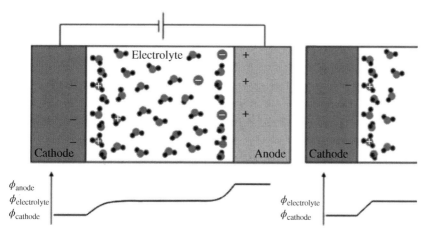

FIGURE 11.2 Illustration of the variation in electrostatic potential in an electrochemical cell. Since the potential becomes constant in the electrolyte, one can treat the two electrodes independently.

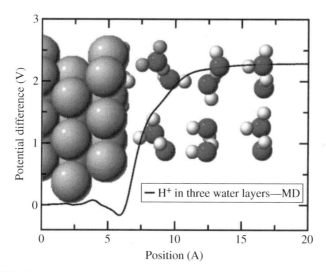

FIGURE 11.3 The electrostatic potential (as seen by an electron) outside a charged Pt(111) slab with three water bilayers outside and one solvated hydronium ion (yellow) per unit cell corresponding to a potential of ca. −2 V versus reversible hydrogen electrode (RHE). The electrostatic potential due to the charged interface is averaged parallel to the surface and is calculated from a time average of a density functional theory (DFT) molecular dynamics simulation of a proton solvated in three water layers obtained after equilibration of the system at 300 K. Adapted from Rossmeisl et al. (2008b). (*See insert for color representation of the figure.*)

Figure 11.3 shows the result of a detailed atomistic calculation for a Pt(111) surface at a negative potential such that the concentration of electrons in the surface and protons in the immediate vicinity is 1 per 6 Pt atoms. It can be seen that the width of the dipole layer is only a few Å—the protons tend to reside in the first water layer.

(a) (b)

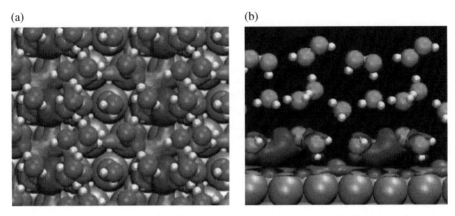

FIGURE 11.4 (a) Top view and (b) side view of a solvated protons in three water layers on top of a Pt(111) electrode. The blue isosurfaces are regions of positive charge around the proton solvated in the water. The purple isosurfaces on the Pt surface are regions of negative charge at the electrode surface. In this case, the proton concentration is very high (one proton per six surface atoms corresponding to a potential of ca. $-2\,V$ vs. RHE). Taken from Skulason et al. (2010) with permission from The American Chemical Society. (*See insert for color representation of the figure.*)

The potential is highly inhomogeneous parallel to the surface. This is illustrated by the charge density plots in Figure 11.4.

11.2 ELECTRON TRANSFER PROCESSES AT SURFACES

Elementary surface reactions at an electrode involve electron transfer processes. Take as an example the last step in the formation of water in the oxygen reduction reaction (ORR) (or, in the reverse direction, the first step in the water splitting reaction):

$$*OH + H^+(aq) + e^-(U) \rightarrow H_2O(aq) + * \qquad (R11.3)$$

The asterisk, *, signifies an adsorption site on the surface, (aq) a species solvated in the aqueous electrolyte, and (U) an electron at a potential U in the electrode.

Below, we will discuss five important ways in which elementary surface electrochemical reactions may differ from their gas–surface counterparts:

1. The chemical potential of the electrons entering the reaction is controlled by the potential of the electrode. This is by far the largest effect. Changing the potential by 1 V changes the reaction free energy of a reaction like R11.3 by 1 eV. This could only be achieved in the equivalent gas-phase reaction using gas-phase hydrogen by changing the pressure of H_2 by ca. 34 orders of magnitude at 300 K.

2. The surface species, as well as the reactants and products, will be solvated by the electrolyte. Reactant and product solvation in the bulk electrolyte are the same as

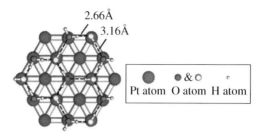

FIGURE 11.5   Structure of coadsorbed water and OH on Pt(111), determined by low-energy electron diffraction (LEED), X-ray photoelectron spectroscopy (XPS), X-ray absorption spectroscopy (XAS), and Auger electron spectroscopy (AES), in combination with DFT calculations. From Schiros et al. (2007).

is in ordinary liquid-phase chemistry. Solvation effects are large and many of them have been studied and tabulated. Solvation effects of species adsorbed on the surface are generally smaller than in the bulk of the electrolyte, since the water only has access to part of the adsorbed molecule and the geometry is strongly restricted by the surface and the bonding of the adsorbate to it. Some of the strongest effects are related to hydrogen bonding of adsorbed *OH groups. Adsorbed *OH on a Pt(111) surface is stabilized on the order of 0.5 eV by the surrounding water. This is due to a very strong hydrogen bonding network, which is formed by a structure with one-third coverage of OH and one-third coverage of H_2O, shown in Figure 11.5. Atomic adsorbates such as O or H that are bound close to the surface show much smaller effects, typically less than 0.1 eV.

3. Solvation events can contribute to activation free energies during transfer of molecules or ions to or from the surface. Since the solvent molecules at the surface are constrained by the presence of the surface, they respond slowly to perturbations, and the processes need not be adiabatic with respect to the solvent degrees of freedom. This could result in "apparent" activation free energies that are larger than the ones you would get if the solvent could relax completely during the process.

4. The electric field at the solid–electrolyte interface will change the adsorption energy. As discussed in Chapter 10, the interaction between an electric field and an adsorbate can be expanded in powers of the electric field. The linear term has a coefficient given by the dipole moment of the adsorbed state, and the second-order term has a strength given by the polarizability. This is illustrated for intermediates of the water splitting (and the reverse oxygen reduction) reaction and for adsorbed CO_2 in Figure 11.6. As shown in Figure 11.1, field strengths outside the surface can be quite high, easily several times 0.1 V/Å. Even so, the effect on the interaction energy with the surface for both adsorbed states and transition states is modest. Exceptions will include adsorbates with very large dipole moments, such as large molecules with a redox center far from the surface or molecules with a very large polarizability. The latter is illustrated by the adsorption of CO_2,

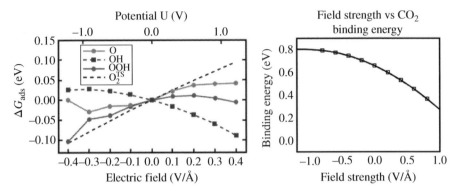

FIGURE 11.6   Variation in the adsorption energy with electrical field strength outside a Pt(111) surface. Left: intermediates in the water splitting reaction and ORR. Right: adsorbed CO_2. Adapted from Karlberg et al. (2007) and Shi et al. (2013).

where positive fields (negatively charged surfaces) result in the formation of a CO_2^--like state outside the surface.

Note that since the electric field outside of a surface is far from uniform (see Fig. 11.4), the electric field effect is not simply proportional to the potential. At potentials low enough, the perturbed regions do not overlap. Adsorbates and processes in the vicinity of an ion outside the surface will be perturbed in much the same way, irrespective of the overall potential. A different potential then primarily means that there are more ions close to the surface and a larger fraction of the surface is perturbed.

5. The electron transfer itself can be rate limiting. This is often the case in electron transfer processes in solution, but for processes taking place close to metal surfaces, the electron transfer rate is usually very high and not rate limiting. For example, in Figure 8.4, the width, Δ, of the adsorbate-induced states outside metal surfaces is of the order of eV. Through the Heisenberg uncertainty principle, $\Delta \tau > \hbar/2$, we get that the lifetime, τ, of electrons on the adsorbate is of the order of 10^{-15} s or, equivalently, that the rate of electron jumps, $1/\tau$, between the surface and the adsorbate is $\sim 10^{15}$ s^{-1}. The charge transfer to completely occupy the bonding states between the oxygen 2p states and the metal states (formally making it an adsorbed screened O^{2-} ion) is therefore instantaneous on the timescale of the proton transfer.

One could imagine processes where the charge transfer happens further from the surface (e.g., to a large molecule where the redox center is buried deeply), where charge transfer rates may have to be considered. A case where charge transfer will definitely have to be taken into account is when the catalyst is an insulator; here, electrons need to tunnel (or hop via defects) through the bandgap from the underlying electrode to the surface. When the electrons (or holes) are created inside the insulator through photon absorption, transport occurs not through tunneling, but through conduction in the bands or the hopping of polarons (localized electrons and holes perturbing the lattice of the insulator). In addition, the rate may be limited by recombination of electrons and holes.

11.3   THE HYDROGEN ELECTRODE

Before discussing the energy diagrams for electrochemical surface reactions, we introduce a very useful concept from electrochemistry, the normal hydrogen electrode (NHE) and the related RHE. The convention is to use the equilibrium between protons and electrons at a given potential with gas-phase H_2 to define a reference potential so that at $U = 0$ V with respect to the NHE, the following reaction is in equilibrium at standard conditions ($p = 1$ bar, $T = 298.15$ K, pH $= 0$):

$$H^+(aq) + e^-(U) \rightleftarrows \frac{1}{2} H_2 \tag{R11.4}$$

The RHE defines the equilibrium potential of R11.4 to be 0 at any pH. It differs from the NHE scale by the free energy difference between pH $= 0$ and any pH:

$$e^{-1} kT \ln c_{H^+} = -e^{-1} kT\ 2.30\ \text{pH}$$

The hydrogen electrode provides a direct link between the free energies in gas-phase adsorption and those relevant in electrochemistry.

The reaction free energy for R11.3, for instance, becomes

$$\Delta G(U_{RHE}) = \Delta G_g + eU_{RHE} + G_{interface} \tag{11.1}$$

at potential U_{RHE} versus RHE. Here, ΔG_g is the free energy difference of the equivalent gas-phase reaction ($*OH + \frac{1}{2} H_2 \rightarrow H_2O(aq) + *$)

$$\Delta G_g = \Delta G° - TS_{ads}^{conf} = \Delta E - T\Delta S° - TS_{ads}^{conf} \tag{11.2}$$

at standard conditions (see Chapter 3), ΔE is the energy difference (including zero-point energy contributions), and S_{ads}^{conf} is the usual configurational entropy of the adsorbed state (Eq. 3.25). $G_{interface}$, is the energy difference between adsorbed OH at the gas–solid and the electrolyte–solid interface due to solvent and field effects (points 2 and 4 in the earlier discussion). In the following, we will skip the RHE subscript unless explicitly needed. This relationship can be generalized for any elementary reaction step involving a single electron and proton transfer and, indeed, to any reaction involving a coupled electron–ion transfer.

11.4   ADSORPTION EQUILIBRIA AT THE ELECTRIFIED SURFACE–ELECTROLYTE INTERFACE

Surface electrochemistry has a direct analogy to the adsorption isotherm in surface chemistry. In surface chemistry, the chemical potential is varied with pressure, while in electrochemistry, the chemical potential can also be varied through potential, which provides a much more powerful handle. Figure 11.7 shows the coverage of

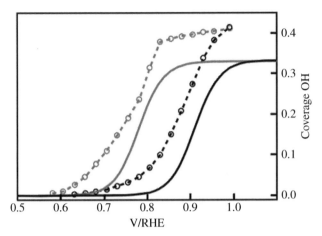

FIGURE 11.7   Experimental (open circles/dashed line) and calculated (full line) coverage of OH versus potential (vs. RHE) for water splitting over Pt(111) (light gray) and Pt$_3$Ni (black). Experimental data are from Stamenkovic et al. (2007), and the theoretical data are from Rossmeisl et al. (2008a). Adapted from Rossmeisl et al. (2008a).

OH as a function of potential versus RHE for two different surfaces, Pt(111) and Pt$_3$Ni(111). The adsorption reaction is R11.3, and the equilibrium condition is

$$\Delta G\left(U,\theta_{OH}\right)=0,$$

which determines the OH coverage versus potential. The experimental data are compared to a simple model based on the calculated adsorption energy of OH (in the presence of water, ~−0.8 eV relative to H$_2$O and H$_2$ up to a coverage of 1/3) on Pt(111), using Equation (A.3.3.5) to evaluate the configurational entropy. Data for a Pt overlayer on Pt$_3$Ni are included, and it is seen that the onset of dissociation of water is shifted due to a weaker surface–OH bond on Pt$_3$Ni. It illustrates that the shifts in adsorption energy associated with d-band shifts (see e.g. Fig. 8.18) are equally important for electrochemistry.

11.5   ACTIVATION ENERGIES IN SURFACE ELECTRON TRANSFER REACTIONS

For elementary processes involving charge transfer at the surface, one can define a potential energy diagram for the process completely as for other surface processes. Since the energy of the electron entering is now potential-dependent, it means that the potential energy diagram becomes potential-dependent.

Consider again the elementary reaction in R11.3. Figure 11.8 shows a calculated potential energy diagram at three different potentials for the proton transfer from the water layer to an adsorbed OH group on the surface.

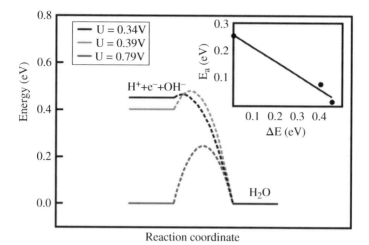

FIGURE 11.8 Potential energy profiles for reducing one OH in the half-dissociated water network by a proton from the water layer at three different potentials. Inset shows the Brønsted–Evans–Polanyi (BEP) relationship for the charge transfer process. The line represents a fit to the data showing a transfer coefficient (γ in Eq. 6.5) of 0.5. Adapted from Tripkovic et al. (2010). (*See insert for color representation of the figure.*)

A noticeable feature of Figure 11.8 is that the activation barrier is always small; the maximum is of the order of 0.25 eV. This is comparable to proton transfer activation energies in water, suggesting that the proton transfer process to an adsorbed OH is not so different from those encountered in proton hopping in water.

Figure 11.8 illustrates an important property of proton transfer reactions at a surface. There is a monotonic relationship between the activation energy and the reaction energy. This is just another example of an activation (or transition state) energy scaling relation, which we have already encountered for other surface reactions in Chapter 6. The transition state energy scales not only with surface binding energies but also with electrode potential. Such relationships are used extensively in electrochemistry. Using the nomenclature from Equation (6.5),

$$E_a = \gamma \Delta E + \xi, \tag{11.3}$$

where the slope, γ, is called the transfer coefficient and a value of 0.5 is often used. Figure 11.8 shows an example where this is actually the case.

As would be expected, the usual scaling relations found in gas-phase adsorption are also found at the solid–liquid interface. Figure 11.9a shows that the scaling line may shift due to solvation of the adsorbates by water, while Figure 11.9b and c shows that similar scaling relations exist for oxide surfaces.

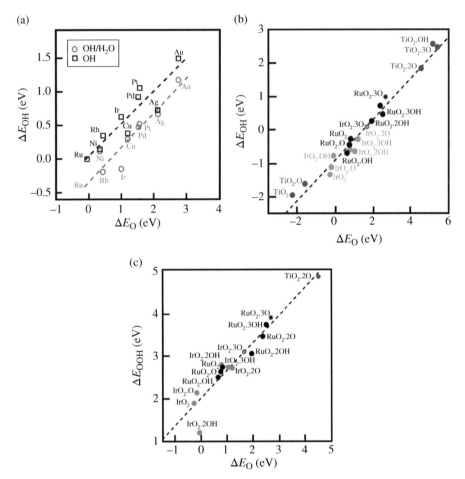

FIGURE 11.9 Scaling relations between adsorption energies of OH and OOH versus the O adsorption energy. (a) data for close-packed metal surfaces. Results where the solvation by H_2O is included are compared to results with no solvation. (b) and (c) data of transition metal oxide surfaces. Adapted from Karlberg and Wahnström (2005) and Rossmeisl et al. (2007).

11.6   THE POTENTIAL DEPENDENCE OF THE RATE

Equation (11.3) has important consequences. Since the reaction energy, ΔE, for a reaction like R11.1 must depend on the potential as

$$\Delta E = \Delta E\left(U=0\right)+eU = \Delta E_0 + eU, \tag{11.4}$$

the activation energy depends on the potential as

$$E_a(U) = \gamma eU + \left(\gamma\Delta E_0 + \xi\right) = \gamma eU + E_{a0} \tag{11.5}$$

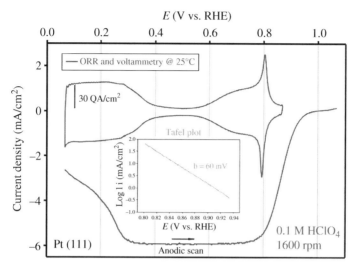

FIGURE 11.10 Measured polarization curve for the ORR over Pt(111). The insert shows the Tafel plot of log current versus potential. From the slope, a transfer coefficient of the order 1 can be extracted $(b \equiv [d\log(r)/dU]^{-1} \cong 60\,\text{mV}/\gamma)$. Figure design adapted from Markovic et al. (1995).

If we neglect variations in the coverage of intermediates with potential (we will return to this question in the following text), the rate of this elementary step will vary with potential as

$$r(U) = \upsilon\, e^{-E_a(U)/kT} = A e^{-\gamma eU/kT}, \tag{11.6}$$

which is the Tafel equation. A plot of the logarithm of the rate versus U should therefore give a linear plot—the Tafel plot—with slope γ.

Figure 11.10 includes an experimental Tafel plot for the ORR ($O_2 + 4(e^- + H^+)$ $\to 2H_2O$). The Tafel slope in the experimental data suggests a transfer coefficient of ca. 1. Had the water formation step, R11.3, been solely responsible for the rate of the full reaction at the potentials of the experiment, it would have been 0.5, according to Figure 11.8.

The transfer coefficient can be understood from the full free energy diagram for oxygen reduction, shown in Figure 11.11. The highest transition state free energies are associated with the activation of O_2 on the surface. The rate of this step is given by

$$r = k_{ads} x_{O_2} \theta_*$$

where k_{ads} is a rate constant and x_{O_2} is the concentration of oxygen molecules in solution. Figure 11.7 shows that in the potential region 0.8–1 V, the surface is mainly covered by *OH, and $\theta_* \simeq 1 - \theta_{OH}$. The transfer coefficient of 1 comes from the fact that for low values of θ_* we have

$$\theta_* \simeq e^{-\Delta G_{OH \to H_2O}/kT} \propto e^{-eU/kT}$$

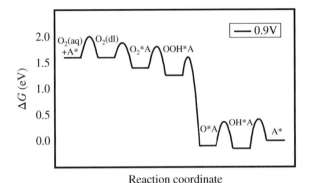

FIGURE 11.11 Calculated free energy diagram for the full ORR over a Pt(111) surface at $U=0.9\,V$. The different elementary reaction steps included are as follows, in order from left to right: diffusion of O_2 from the bulk electrolyte to the region (double layer) just outside the surface (the effective free energy barrier shown is deduced from the diffusion rate); adsorption of O_2 (note that this involves electron transfer to the O_2 molecule, but not a whole electron, and the electron transfer is there also in the absence of the potential since a metal surface has a large pool of electrons available at the Fermi level), followed by four coupled electron–proton transfers to form water; and recreation of the adsorption site *A. Adapted from Hansen et al. (2014).

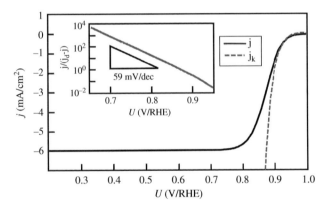

FIGURE 11.12 Calculated polarization curve for the ORR over Pt(111). The insert shows the Tafel plot of log current versus potential. The black curve includes diffusion limitations present in the experiment, while the solid curve is the kinetic current density in the absence of such limitations. Adapted from Hansen et al. (2014).

k_{ads} is, to a first approximation, independent of the potential, since it is mainly related to rearrangement of the water outside the surface during adsorption, and an effective transfer coefficient of 1 for the full reaction results.

Figure 11.12 shows the calculated polarization curve (current density vs. potential) and the Tafel plot based on the potential energy diagram in Figure 11.11 to be compared to the experimental data in Figure 11.10.

11.7   THE OVERPOTENTIAL IN ELECTROCATALYTIC PROCESSES

Free energy diagrams for full electrocatalytic reactions, like in Figure 11.11, are as useful in understanding surface electrocatalysis as those introduced for heterogeneous catalysis in Chapter 3. Since energy barriers are quite small for proton transfer reactions and since the barrier scales with the reaction energy (as illustrated in Fig. 11.8), it is often useful to consider simplified free energy diagrams where only the energies of intermediates are included. The potential dependence of the reaction is easily seen by showing the variation of the free energy diagram with potential. Figure 11.13 shows such plot for the oxygen evolution reaction (OER) and, in the reverse direction, ORR.

Experimental data for the variation of the rate of the OER and ORR with potential for a number of materials are shown in Figure 11.14. For all the electrocatalysts, the OER does not start to have an appreciable current density j until a substantial overpotential

$$\eta(j) = |U(j) - U_{eq}| \qquad (11.7)$$

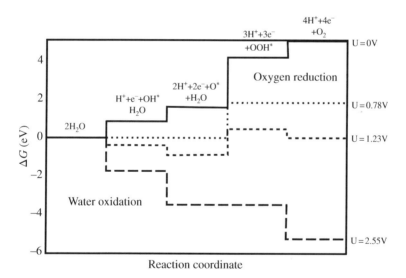

FIGURE 11.13 Free energy diagram calculated for a Pt(111) surface for the water oxidation, OER from left to right and the ORR from right to left. The free energy for each intermediate is shown at different potentials versus RHE. At zero potential, OER is very uphill in energy, while the ORR is very downhill. At the equilibrium potential for the reactions, $U_{eq} = 1.23\,V$, there are substantial energy barriers separating initial and final states in both directions, making them slow. The ORR is only downhill in free energy for potentials up to 0.78 V, while a potential of 2.55 V is needed for OER to be all downhill on Pt(111). At potentials above ca. 0.9 V, Pt(111) oxidizes so the calculation for a metallic Pt(111) surface is purely hypothetical and to illustrate the principle. Adapted from Rossmeisl et al. (2005).

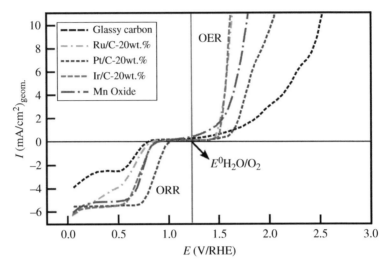

FIGURE 11.14 Measured current density as a function of applied potential for the ORR and OER over different electrode catalysts. $U_{eq} = E^0_{H_2O/O_2} = 1.23\,V$ is indicated. Adapted from Gorlin and Jaramillo (2010).

is applied. Typically, the best OER (RuO_2 and IrO_2) and ORR (Pt) catalysts only give current densities above $|j| = 5\,mA/cm^2$ at potentials that are $|\eta| \sim 0.3\,V$ above or below the equilibrium potential, $U_{eq} = 1.23\,V$.

In Figure 11.13, the energy of each intermediate varies with the potential, as shown in Equation (11.4). The ORR over Pt is only downhill in free energy for potentials above ~0.8 V, ca. 0.4 V below U_{eq}. The OER is only exergonic for all reaction steps at a potential of 2.55 V. While the former is in good agreement with the measured overpotential at current densities of the order of $5\,mA/cm^2$, the latter is much higher than measured. The reason is that under the high potentials needed for OER, Pt is oxidized, and PtO_2 has a considerably lower overpotential (a similar calculation on PtO_2 gives a value of 1.7 V, close to the experimental result in Fig. 11.14).

The previous examples can be generalized as follows. According to Equation (R11.4), we can write the free energy change for any elementary step, i, in an electrochemical reaction with a transfer of one electron and a proton as

$$\Delta G_i(U) = \Delta G_{0,i} \pm eU, \tag{11.8}$$

where the sign of the last term depends on whether the electron transfer is from or to the surface. If the ion transferred is not a proton, a similar expression can be obtained with a potential reference given by the chemical potential of this ion in the system. We can now define the limiting potential for elementary reaction step i as the potential where the free energy difference for the reaction is zero:

$$U_{L,i} = \frac{\mp \Delta G_{0,i}}{e} \tag{11.9}$$

Consider a reduction reaction like the ORR, where electrons are transferred from the surface to the reactant (this corresponds to a + sign in Equation (11.8) and a − sign in Equation (11.9)). The minimum of the $U_{L,i}$s for the elementary steps making up the reaction defines the potential where all steps are exergonic, and this potential is termed the limiting potential for the reaction

$$U_{L,\text{red}} = \min\{U_{L,i}\} \tag{11.10}$$

We define the theoretical overpotential as

$$\eta_{\text{theo,red}} = U_{\text{eq}} - U_{L,\text{red}} \tag{11.11}$$

Likewise, for an oxidation reaction like the OER, the maximum value of $U_{L,i}$ for which all reaction steps are exergonic will define the limiting potential for the full reaction:

$$U_{L,\text{ox}} = \max\{U_{L,i}\} \tag{11.12}$$

and we define the theoretical overpotential as

$$\eta_{\text{theo,ox}} = U_{L,\text{ox}} - U_{\text{eq}} \tag{11.13}$$

Note that this is not the same as the measured overpotential (Eq. 11.7), which depends on the current density. However, η_{theo} is a measure of the activity and has been found in a number of cases to scale with the measured overpotential at a fixed current density for different catalysts.

11.8 TRENDS IN ELECTROCATALYTIC ACTIVITY: THE LIMITING POTENTIAL MAP

Because of the scaling relations, each of the reaction free energies in the ORR will scale with the O or the OH adsorption energy. That means that we can define the limiting potential for each step as a function of the OH adsorption energy:

$$U_{L,i} = U_{L,i}(\Delta G_{\text{OH}})$$

Figure 11.15 shows the limiting potential for the two elementary reaction steps of the ORR with the lowest limiting potential as a function of ΔG_{OH}. The lowest limiting potential for each OH adsorption energy defines the total limiting potential (Eq. 11.10), and defines a limiting potential volcano.

For strong binding energies of OH, the lowest value of $U_{L,i}$, defining the left leg of the volcano, is that for the removal of adsorbed OH ($*\text{OH}+\text{H}^++\text{e}^- \rightarrow \text{H}_2\text{O} +*$). The stronger the OH is bound to the surface, the more difficult it is to remove it. The right leg is given by the process of activating molecular O_2. A weak surface–oxygen bond

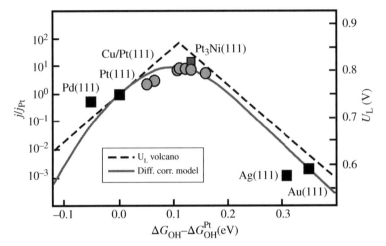

FIGURE 11.15 Dashed lines indicate the limiting potential as a function of the OH adsorption energy, while the full line is the kinetic activity map at 0.9 V versus RHE based on the free energy diagram in Figure 11.12. Experimental data for (111) facets measured at the same potential are included. All data are shown relative to Pt(111). Theoretical values are from Hansen et al. (2014). Experiments labeled Cu/Pt(111) are Pt overlayers on CuPt near-surface alloys from Stephens et al. (2011). Other experimental data are from Blizanac et al. (2004); Wang et al. (2004) for Pt(111); Shao et al. (2006) for Pd(111); Blizanac et al. (2006) for Ag(111) and Ag(100); Stamenkovic et al. (2007), 315, 493 for Pt$_3$Ni(111). Adapted from Hansen et al. (2014).

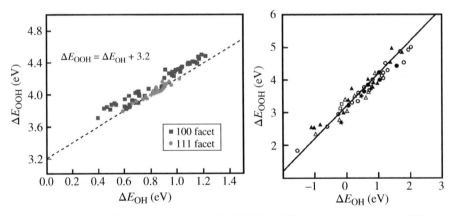

FIGURE 11.16 Scaling relation between the OOH and OH adsorption energy on different metals (left) and metal oxides (right). Adapted from Viswanathan et al. (2012) and Man et al. (2011).

makes the formation of adsorbed OOH more difficult, and hence, this step becomes potential limiting. Pt is close to the top but not at the top. It binds OH (and O) a little too strongly. There are alloys of Pt that are closer to the top, but it is clear from this analysis that the value of the maximum still corresponds to a sizable overpotential.

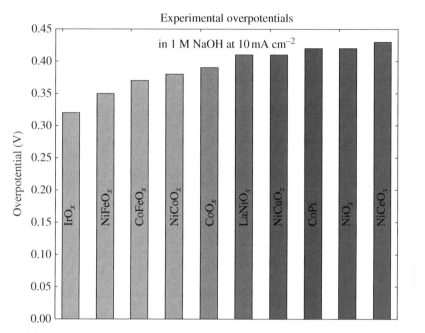

FIGURE 11.17   Measured OER overpotential for a series of catalysts under similar conditions. Adapted from McCrory et al. (2013).

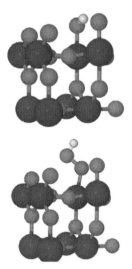

FIGURE 11.18   Structure of adsorbed OH and OOH on $RuO_2(110)$.

This explains why it has, to date, been impossible to find electrode catalysts for the ORR with a really low overpotential.

In Figure 11.15, we also include the result of a full kinetic model giving current densities at $U = 0.9\,V$ as a function of the OH adsorption energy. The figure also includes experimentally measured current densities. Clearly, both the full kinetic model and the much simpler limiting potential analysis describe trends very well, and the latter provides a useful way to understand trends in electrocatalytic activity.

The position and value of the top of the limiting potential map (and the kinetic activity map) are given by the relative positions of the two limiting potential lines defining the volcano. That, in turn, is given by the scaling relation between the OOH and OH adsorption energies (Fig. 11.16). It turns out that all the metals and their oxides give a relationship between the OH and OOH adsorption energy of

$$\Delta G_{OOH} = \Delta G_{OH} + 3.2(\pm 0.2) \text{ eV} \tag{11.14}$$

Since there are two electron–proton transfers separating *OOH and *OH, even the best catalyst on this line will need $3.2\,eV/2 = 1.6\,eV$ per electron transfer. That is ca. $0.4\,eV$ from the equilibrium potential for the reaction and suggests that any materials obeying Equation (11.14) will have a minimum theoretical overpotential of this size. This is consistent with experimental overpotentials observed for ORR (Fig. 11.15) and for the OER (Fig. 11.17).

While it is important to optimize materials along the scaling line—by alloying Pt, for instance, or by increasing the number of active sites and their stability—real breakthroughs in the search for new ORR and OER electrocatalysts must come from new classes of materials that break the scaling. This may be difficult with planar surfaces with only a single type of active site, but it should be possible to engineer more interesting surfaces. We need to find ways of stabilizing *OOH (or *OH and *O) relative to *OH. Figure 11.18 shows the structure of the two adsorbates on RuO_2, coadsorbates that interact more strongly with the longer OOH molecule than OH could lead to a break in the scaling between them.

REFERENCES

Blizanac BB, Lucas CA, Gallagher ME, Arenz M, Ross PN, Markovic NM. Anion adsorption, CO oxidation, and oxygen reduction reaction on a Au(100) surface: the pH effect. J Phys Chem B 2004;108:625.

Blizanac BB, Ross PN, Marković NM. Oxygen reduction on silver low-index single-crystal surfaces in alkaline solution: rotating ring disk(Ag(hkl)) studies. J Phys Chem B 2006; 110:4735.

Gorlin Y, Jaramillo TF. A bifunctional nonprecious metal catalyst for oxygen reduction and water oxidation. J Am Chem Soc 2010;132 (13612).

Hansen HA, Viswanathan V, Nørskov JK. Unifying kinetic and thermodynamic analysis of 2 e– and 4 e– reduction of oxygen on metal surfaces. J Phys Chem C 2014;118:6706.

Karlberg GS, Wahnstrom G. An interaction model for OH + H_2O-mixed and pure H_2O overlayers adsorbed on Pt(111). J Chem Phys 2005;122:195705.

Karlberg GS, Rossmeisl J, Nørskov JK. Estimation of electric field on the oxygen reduction reaction based on the density functional theory. Phys Chem Chem Phys 2007;9:5158.

Man IC, Su H-Y, Calle-Vallejo F, Hansen HA, Martínez JI, Inoglu NG, Kitchin J, Jaramillo TF, Nørskov JK, Rossmeisl J. Universality in oxygen evolution electrocatalysis on oxide surfaces. ChemCatChem 2011;3:1159.

Markovic NM, Gasteiger BN, Ross PN. Oxygen reduction on platinum low-index single-crystal surfaces in sulfuric acid solution: rotating ring-Pt(hkl) disk studies. J Phys Chem 1995; 99:3411.

McCrory CL, Jung S, Peters JC, Jaramillo TF. Benchmarking heterogeneous electrocatalysts for the oxygen evolution reaction. J Am Chem Soc 2013;135:16977.

Rossmeisl J, Logadottir A, Nørskov JK. Electrolysis of water on (oxidized) metal surfaces. Chem Phys 2005;319 (178).

Rossmeisl J, Qu ZW, Zhu H, Kroes GJ, Nørskov JK. Electrolysis of water on oxide surfaces. J Electroanal Chem 2007;607:83.

Rossmeisl J, Karlberg GS, Jaramillo TF, Nørskov JK. Steady state oxygen reduction reaction and cyclic voltammetry. Faraday Discuss 2008a;140 (337).

Rossmeisl J, Skulason E, Björketun ME, Tripkovic V, Nørskov JK. Modeling the electrified solid–liquid interface. Chem Phys Lett 2008b;466:68.

Schiros T, Näslund L-Å, Andersson K, Gyllenpalm J, Karlberg GS, Odelius M, Ogasawara H, Pettersson LGM, Nilsson A. Structure and bonding of water-hydroxyl mixed phase on Pt(111). J Phys Chem C 2007;111:15003.

Shao MH, Huang T, Liu P, Zhang J, Sasaki K, Vukmirovic MB, Adzic RR. Palladium monolayer and palladium alloy electrocatalysts for oxygen reduction. Langmuir 2006; 22:10409.

Shi C, O'Grady CP, Peterson AA, Hansen HA, Nørskov JK. Modeling CO_2 reduction on Pt(111). Phys Chem Chem Phys 2013;15:7114.

Skulason E, Tripkovic V, Bjorketun ME, Gudmundsdottir S, Karlberg G, Rossmeisl J, Bligaard T, Jonsson H, Nørskov JK. Modeling the electrochemical hydrogen oxidation and evolution reactions on the basis of density functional theory calculations. J Phys Chem C 2010; 114:18182.

Stamenkovic VR, Fowler B, Mun BS, Wang G, Ross PN, Lucas CA, Markovic NM. Improved oxygen reduction activity on $Pt_3Ni(111)$ via increased surface site availability. Science 2007;315:493.

Stephens IEL, Bondarenko AS, Perez-Alonso FJ, Calle-Vallejo F, Bech L, Johansson TP, Jepsen AK, Frydendal R, Knudsen BP, Rossmeisl J, Chorkendorff I. Tuning the activity of Pt(111) for oxygen electro reduction by subsurface alloying. J Am Chem Soc 2011;133: 5485.

Tripković V, Skúlason E, Siahrostami S, Nørskov JK, Rossmeisl J. The oxygen reduction reaction mechanism on Pt(111) from density functional theory calculations. Electrochim Acta 2010;55:7975.

Viswanathan V, Hansen HA, Rossmeisl J, Nørskov JK. Universality in oxygen reduction electrocatalysis on metal surfaces. ACS Catal 2012;2:1654.

Wang JX, Markovic NM, Adzic RR. Kinetic analysis of oxygen reduction on Pt(111) in acid solutions: intrinsic kinetic parameters and anion adsorption effects. J Phys Chem B 2004; 108:4127.

FURTHER READING

Koper MTM. Thermodynamic theory of multi-electron transfer reactions: implications for electrocatalysis. J Electroanal Chem 2011;660:254.

Liao P, Carter EA. New concepts and modeling strategies to design and evaluate photo-electro-catalysts based on transition metal oxides. Chem Soc Rev 2013;42:240.

Limmer DT, Willard AP, Madden PA, Chandler D. Hydration at a metal surface can be heterogeneous and hydrophobic. Proc Natl Acad Sci 2013;110:4200.

Markovic NM, Ross PN. Surface science studies of model fuel cell electrocatalysts. Surf Sci Rep 2002;45:121.

Nørskov JK, Rossmeisl J, Logadottir A, Lindqvist L, Kitchin JR, Bligaard T, Jónsson H. Origin of the overpotential for the oxygen reduction at a fuel-cell cathode. J Phys Chem B 2004;108:17886.

Peterson AA, Nørskov JK. Activity descriptors for CO_2 electroreduction to methane on transition-metal catalysts. J Phys Chem Lett 2012;3:251.

Suntivich J, Gasteiger HA, Yabuuchi N, Nakanishi H, Goodenough JB, Shao-Horn Y. Design principles for oxygen-reduction activity on perovskite oxide catalysts for fuel cells and metal-air batteries. Nat Chem 2011;3:546.

Wieckowski A, Koper M. *Fuel Cell Catalysis: A Surface Science Approach*. Hoboken, NJ: John Wiley & Sons, Inc; 2009. The Wiley Series on Electrocatalysis and Electrochemistry.

12

RELATION OF ACTIVITY TO SURFACE ELECTRONIC STRUCTURE

As pointed out in the previous chapters, finding the right catalysts for a given reaction is closely linked to understanding the chemical bond formed between the relevant adsorbates and the catalyst. The catalytic cycle always involves an initial adsorption of reactants, which is then followed by either bond rearrangement or diffusion of the reactants. The strength of the bond formed between the adsorbates and the catalyst can provide insights regarding the functionality of the catalyst. As we have seen in previous chapters, an understanding of the bonds formed at the electronic level is essential both for understanding catalytic reactions and for moving toward rational design strategies. In this chapter, we will describe what happens when an adsorbate interacts with the complex electronic structure of a catalyst surface.

12.1 ELECTRONIC STRUCTURE OF SOLIDS

To understand how a given surface (e.g., a metal, semiconductor, or insulator) inter-acts with an adsorbate, we first need to understand the electronic structure of the surface. We also need to understand how interactions between the different constitu-ents in the surface affect its reactivity.

A solid is composed of a large number of atoms that are grouped together in a well-defined crystal structure. Composite states are formed in a solid comprised of the outermost atomic states of each individual atom. What happens is that the energy

Fundamental Concepts in Heterogeneous Catalysis, First Edition. Jens K. Nørskov,
Felix Studt, Frank Abild-Pedersen and Thomas Bligaard.
© 2014 John Wiley & Sons, Inc. Published 2014 by John Wiley & Sons, Inc.

levels are split into bands of energy levels and the states will be delocalized, meaning that they will no longer be associated with one atom alone. In order to understand these interactions, let us consider a system consisting of just two atoms.

When two separated atoms (with well-defined eigenstates and eigenenergies) are brought together, they interact and form a set of composite states. Let ψ_A and ψ_B be normalized eigenfunctions with energy eigenvalues ε_A and ε_B describing the states of two noninteracting atoms A and B. When the two atoms approach each other, they will start interacting via their overlapping molecular orbitals. In a simple form of molecular orbital theory, the eigenfunctions of the composite state are described as a linear combination of the eigenfunctions or orbitals of the individual atoms (LCAO). In the case of the atoms A and B, we can write this combination as

$$\psi_{AB} = c_1\psi_A + c_2\psi_B \tag{12.1}$$

We introduce the overlap matrix elements S_{ij} and the coupling elements V_{ij}

$$S_{ij} = \langle \psi_i | \psi_j \rangle = \begin{cases} \delta_{ij} & i = j \\ S & i \neq j \end{cases} \tag{12.2}$$

$$V_{ij} = \langle \psi_i | H | \psi_j \rangle = V_{ji} = V \tag{12.3}$$

Here and throughout the chapter, we use the bra and ket vector notation as introduced by P. Dirac to describe a specific state vector of a system. Whenever the scalar product of two state vectors, $\langle \psi |$ and $| \varphi \rangle$, is defined, then the complete bracket $\langle \psi | \varphi \rangle = \int \psi^*(r)\varphi(r)dr$ will denote a number.

With the simple ansatz, Equation (12.1), we can write the Schrödinger equation

$$H\psi_{AB} = \varepsilon\psi_{AB} \tag{12.4}$$

in matrix form

$$\left(\overline{\overline{H}} - \overline{\overline{S}}\varepsilon\right)\overline{c} = 0 \tag{12.5}$$

with solutions or the energy eigenvalues

$$\varepsilon_{\pm} = \frac{(\varepsilon_A + \varepsilon_B) - 2SV \mp \sqrt{(\varepsilon_A - \varepsilon_B)^2 - 4(\varepsilon_A + \varepsilon_B)SV + 4V^2 + 4\varepsilon_A\varepsilon_B S^2}}{2(1 - S^2)} \tag{12.6}$$

Typically, the overlap between states is small compared to the other quantities in the square root. Thus, to a first-order approximation in S, we can simplify this equation to give

$$\varepsilon_{\pm} = \frac{(\varepsilon_A + \varepsilon_B) - 2SV \mp \sqrt{(\varepsilon_A - \varepsilon_B)^2 + 4V^2}}{2} \tag{12.7}$$

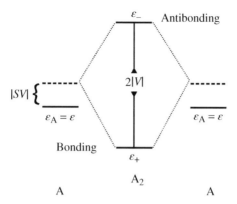

FIGURE 12.1 Schematics showing the interaction between two identical molecular orbitals with energy ε. The interaction introduces repulsion on each level that is proportional to the overlap S, and the subsequent hybridization produces bonding and antibonding states separated by two times the coupling matrix element V.

which can be simplified to give

$$\varepsilon_{\pm} = \frac{(\varepsilon_A + \varepsilon_B)}{2} \mp \sqrt{V^2 + \left(\frac{\varepsilon_A - \varepsilon_B}{2}\right)^2} - SV \qquad (12.8)$$

In the following, we will look at some simple cases of molecular bonds. For a homonuclear system (e.g., He–He, H–H, N–N, etc.), the expression for the energy levels is especially simple. Since in this case $\varepsilon_A = \varepsilon_B$, the energies described by Equation (12.8) reduce to

$$\varepsilon_{\pm} = \varepsilon_A \mp |V| - SV \qquad (12.9)$$

Here, $S>0$ and $V<0$, which shows that both energy levels are shifted up by the repulsive term SV and subsequently they split up in two energy levels separated by two times the coupling matrix element, as seen schematically in Figure 12.1. The upshift in energy by $-SV$ stems from the fact that according to the Pauli principle, the states of the two atoms must be orthogonal to each other. This is known as Pauli repulsion. The formation of two states, one below and one above the average, is a very basic property of a quantum system. The lowest state is referred to as the bonding state, and the high-lying state is the antibonding state.

Let us initially assume that the bond energy between two atoms can be written in terms of differences in the one-electron energies of the occupied orbitals in the molecule and the energy levels of the constituting atoms. Then for two atoms A and B brought together to form a molecule AB, we find that the bond strength is given by

$$\Delta E = \sum_{occ} \varepsilon_i^{AB} - \sum_{occ} \varepsilon_j^{A} - \sum_{occ} \varepsilon_k^{B} \qquad (12.10)$$

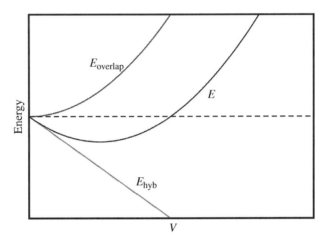

FIGURE 12.2   The binding energy in a homonuclear diatomic molecule as a function of the coupling matrix element. The total energy is split into a repulsive part due to the orthogonalization of atomic states and an attractive term due to hybridization between states.

This is, of course, an oversimplification, since bonding in real systems involves a number of other interactions (e.g., electrostatic interactions), but for a qualitative picture of bonding in molecules, this is a reasonable first approximation. Later, we shall argue in more detail for why this approximation under some circumstances can be very useful.

Clearly, for the molecule to be stable, one requires that the energy of the bonded state be lower than that of the individual atoms (i.e., $\Delta E < 0$). For a molecule like H_2 where we assume that the energy levels of atomic hydrogen possess one electron of energy ε_H, we can calculate the energy difference:

$$\Delta E \simeq 2\varepsilon_+ - 2\varepsilon_H = -2|V| - 2SV \qquad (12.11)$$

From the definition of the coupling elements, V is approximately proportional to the overlap of the eigenfunctions such that $S = -\gamma V$. We can then simplify Equation (12.11) to be

$$\Delta E \simeq -2|V| + 2\gamma V^2 \qquad (12.12)$$

It is seen that the competition between the repulsive term due to the overlapping states $(2\gamma V^2)$ and the attractive hybridization term $(-2|V|)$ will give rise to a chemical bond formed between the two hydrogen atoms. This is illustrated schematically in Figure 12.2.

In the case of two He atoms approaching each other, we have two electrons per energy level. Thus, the bonding and antibonding states will end up being completely filled. If we calculate the one-electron energy associated with this bond formation (Eq. 12.7), we find that the downshift of the bonding state is exactly cancelled out by the upshift of the antibonding state. Hence, only the repulsive term from the orbital overlap is left, meaning that the He$_2$ system is unstable, as expected:

$$\Delta E \simeq 2\varepsilon_+ + 2\varepsilon_- - 4\varepsilon_{He} = -4SV \qquad (12.13)$$

For a heteronuclear bond $\varepsilon_A \neq \varepsilon_B$, the energy levels in the composite system are given by the expression in Equation (12.2). However, in cases where the individual atomic energy levels ε_A and ε_B are well separated relative to the coupling matrix element V, we can Taylor expand Equation (12.7) if we rearrange it slightly:

$$\varepsilon_\pm = \frac{(\varepsilon_A + \varepsilon_B)}{2} \mp \left(\frac{\varepsilon_A - \varepsilon_B}{2}\right)\sqrt{1 + \left(\frac{2V}{\varepsilon_A - \varepsilon_B}\right)^2} - SV \qquad (12.14)$$

This immediately gives to the second order in $V/(\varepsilon_A - \varepsilon_B)$ that

$$\varepsilon_+ \simeq \varepsilon_A - \frac{V^2}{(\varepsilon_B - \varepsilon_A)} - SV \qquad (12.15)$$

$$\varepsilon_- \simeq \varepsilon_B + \frac{V^2}{(\varepsilon_B - \varepsilon_A)} - SV \qquad (12.16)$$

which shows that when energy levels get further and further apart, the splitting of the energy levels due to the interaction decreases. Calculating the bond energy using these approximate values for the energy levels, one finds that

$$\Delta E \simeq -\sqrt{(\varepsilon_B - \varepsilon_A)^2 + 4V^2} - 2SV \qquad (12.17)$$

We note that when $\varepsilon_A \to \varepsilon_B$ or vice versa, the expression for the homonuclear interaction is obtained as expected. In Figure 12.3, we provide an overview of some of the conclusions that can be drawn from molecular orbital theory.

12.2 THE BAND STRUCTURE OF SOLIDS

Solids are often composed of a well-defined array of atoms. Each atom in the array is bound in a potential energy well defined by the surrounding atoms. In the following, we shall consider only solids that are free of defects. Bonding between atoms in a solid can be understood in much the same way as the bonding formed between single molecular orbitals, except that the bonding orbitals are treated as bands of bonding and antibonding orbitals, due to the large number of overlapping energy levels resulting from the large number of atoms in a crystal structure.

A quantitative description of this band structure can be obtained using what is known as a "tight-binding" model.

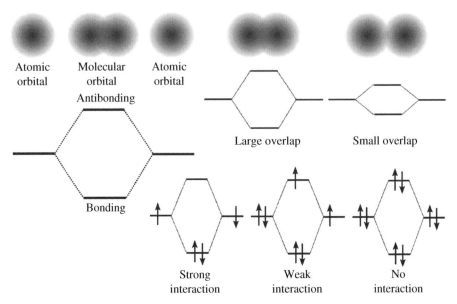

FIGURE 12.3 Schematics of the main conclusions that can be drawn from the LCAO theory. The interaction gives rise to splitting of the atomic orbitals into bonding and antibonding states. The extent of orbital overlap leads to a larger split between the bonding and antibonding energy levels, and the degree of orbital filling determines the strength of the molecular interaction.

In the following, we will describe the structure of solids in more qualitative terms. Let us consider a metal like aluminum to get a feeling for what happens. Aluminum has the electronic configuration $1s^2 2s^2 2p^6 3s^2 3p$, which sometimes is written as $[Ne]3s^2 3p$. This means that aluminum has three electrons circling a neon-like closed electronic shell.

The s and p electronic states have extended electronic orbitals, and as we saw in the molecular orbital approach, this will give large overlaps between electronic states and hence large splitting of the bonding and antibonding states. In metals, the atomic orbitals are many, and in Figure 12.4, we have shown qualitatively how a large number of strongly interacting states give rise to a broad continuous band of states. This is identical to saying that the s and p electrons are not restricted to specific atoms in the lattice and hence are strongly delocalized.

In catalysis, the interest is mainly focused on the transition metals. They distinguish themselves from other metals by having partially or completely filled d-shells. The orbitals of d electrons have specific shapes and very localized spatial extent. Hence, the overlap between d electronic states is much smaller than for s and p electronic states. When the interaction is weak, one still gets a continuum of states that form a band structure, but the band is much narrower.

In Figure 12.5, we show the band structure for two electronically very different metals: aluminum (Al) and silver (Ag). The electronic configuration for Ag is $[Kr]5s4d^{10}$ showing that it has 11 electrons outside a krypton-like closed shell.

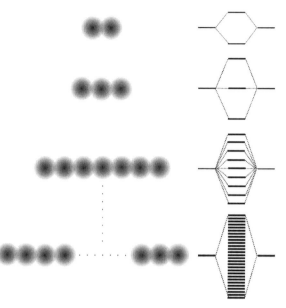

FIGURE 12.4 Schematics showing what happens in the limit of infinitely many overlapping orbitals.

The bands have been calculated using *ab initio* density functional theory (DFT) with 3 valence electrons and 11 valence electrons for Al and Ag, respectively. Each distinct energy level in the band is called a state; the number of states with a given energy, the density of states (DOS), is shown in Figure 12.5. For aluminum, we see the features of the *s*- and *p*-bands and how they extend over the entire energy region. The projected DOS shows the atomic level contributions to the band structure, and it reveals how delocalized the electronic states are in aluminum.

The behavior of the electrons in Al resembles the behavior of a gas of free electrons in three dimensions. To see this, we recall the energy solutions derived in basic quantum mechanics for a free electron gas:

$$\varepsilon_k = \frac{\hbar^2 k^2}{2m} \tag{12.18}$$

where k is a wave vector characterizing the state ($p = \hbar k$ is the momentum of the electron in state k). If we let N be the total number of orbitals within a sphere in k-space of volume $V_s = 4\pi k^3/3$ and then using that there is one allowed k-vector per spin state within the volume element $(2\pi/L)^3$, we can solve for N to find $N = V/3\pi^2(2m\varepsilon/\hbar^2)^{3/2}$. This immediately provides an expression for the DOS for a free electron gas in three dimensions as

$$\text{DOS} = \frac{dN}{d\varepsilon} = \frac{V_s}{2\pi^2} \left(\frac{2m}{\hbar^2}\right)^{3/2} \varepsilon^{1/2} \tag{12.19}$$

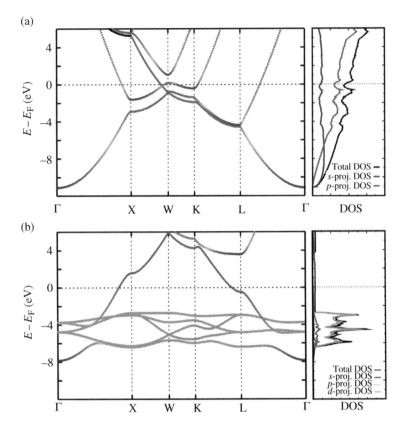

FIGURE 12.5 Calculated band diagrams and corresponding densities of states for the face-centered cubic transition metals (a) aluminum (Al) and (b) silver (Ag). The bands are plotted along high symmetry directions of the reciprocal lattice in k-space.

If we compare this result with the DOS for aluminum shown in Figure 12.5a, one sees that the total density of occupied electronic states has exactly a square root dependence on the energy.

For Ag, we again see the broad features from the delocalized sp electrons overlapping in the crystal lattice, but more importantly, we see a large number of almost flat bands in the -3 to -6 eV region. These flat bands originate from the localized (spatially confined) d electrons. This leads to a clustering of states in this region as seen in the DOS in Figure 12.5b.

What is important to realize from this section is that interacting valence electrons give rise to a band of states due to the splitting of energy levels into bonding and antibonding states. The width of the band is proportional to the strength of the coupling, and the coupling is a function of distance and number of neighbors. This means that if a crystal lattice is expanded or compressed it will result in a narrowing or broadening of states, respectively.

Besides metals, which all have states crossing the Fermi level, there are also systems that have bands that are even more complex than bands in metals. Depending on

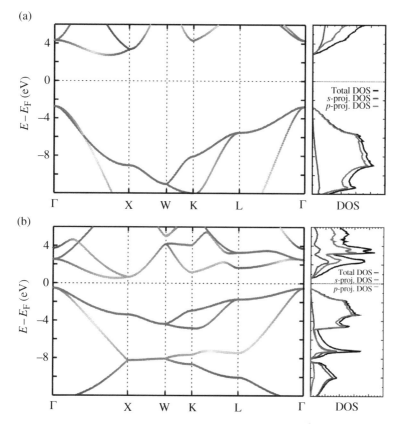

FIGURE 12.6   Calculated band diagrams and corresponding densities of states for (a) bulk diamond, which is an insulator, and (b) bulk Si, which is a semiconductor. The bands are plotted along high symmetry directions of the reciprocal space for the two crystal structures.

how the different orbitals in the material hybridize with each other, band crossings and even band openings can be formed in the near Fermi-level region. Due to this, a variety of different materials with different properties emerge, such as conductors, semiconductors, or insulators.

A material is considered a good conductor if electrons can be excited from the valence band below the Fermi level to the conduction band above the Fermi level, where the electron can move around freely in the material. We note that very small bandgap or zero bandgap materials are the only materials that make good conductors. Hence, materials with a free electron metal band structure with DOS described by Equation (12.19) are good conductors, and transition metals that have similar sp-bands as the free electron metals also make good conductors.

The other limit is where the bandgap formed is of such size that it is nearly impossible to excite electrons into the conduction band. As a rule of thumb, materials are considered insulators when the bandgap is larger than 5 eV. In Figure 12.6a, we have shown the band structure and the corresponding projected DOS for diamond, which is a known insulator. Our calculations also show that the bandgap is more than $5\,\mathrm{eV}$, which is in good agreement with the experimental value of $5.41\,\mathrm{eV}$.

A semiconductor is a material that behaves exactly like an insulator at $T=0\ K$. However, the energy needed for the electrons to jump from the top of the valence band to the bottom of the conduction band is such that at nonzero temperature it increases appreciably. Clearly, the energy required to make this jump will depend on the material and whether there are defects, impurities, or dopants in the material. The transition can occur when the materials either absorb a phonon (heat-induced transition) or a photon (light-induced transition). These features make semiconductors very important materials in electronic devices because carrier densities can be easily tuned by controlling dopant levels. In Figure 12.6b, the calculated band structure and DOS for silicon are shown. The bandgap is close to $1\,\mathrm{eV}$ for this material. This is in reasonable agreement with the experimental value of $1.14\,\mathrm{eV}$, thus showing that the electrons in Si can be transferred into the conduction band easily.

So far, we have focused on describing the structure of the materials upon which our catalytic reactions are supposed to take place. The next step is to look at what happens when a molecule approaches the surface of such a material and begins to interact with the material and hence forms a chemical bond. The Newns–Anderson model is a model that describes the hybridization of a single adsorption state on an atom or a molecule with the large continuum of states at the surface.

In the following, we will describe the Newns–Anderson model in detail.

12.3   THE NEWNS–ANDERSON MODEL

Consider a metal surface with one-electron states $|k\rangle$ with energies ε_k, and an adsorbate with a single valence state $|a\rangle$ of energy ε_a. When the adsorbate approaches the surface from far away to a position just outside, the two sets of states are coupled by matrix elements $V_{ak} = \langle a|H|k\rangle$, where H is the Hamiltonian of the combined system. If we expand the solutions $|\varphi_i\rangle$ of H in terms of the free adsorbate and surface solutions

$$\varphi_i\rangle = c_{ai}\,|\,a\rangle + \sum_k c_{ki}\,|\,k\rangle, \tag{12.20}$$

and neglect the overlap $\langle a|k\rangle$, then the Schrödinger equation can be written as

$$H\overline{c}_i = \varepsilon_i \overline{c}_i, \tag{12.21}$$

where $H_{aa} = \varepsilon_a$, $H_{kk} = \varepsilon_k$, and $H_{ak} = V_{ak}$.

The projection of the DOS on the adsorbate state can be written as

$$n_a(\varepsilon) = \sum_i \left|\langle \varphi_i\,|\,a\rangle\right|^2 \delta(\varepsilon - \varepsilon_i), \tag{12.22}$$

where the sum is over the eigenstates of the full Hamiltonian.

It is more advantageous to consider this quantity since it maps out the evolution of the original adsorbate state as it approaches the surface and begins interacting with the surface metal states.

Using the fact that a Lorentzian becomes a delta function in the limit $\delta \to 0^+$, we can rewrite this as

$$n_a(\varepsilon) = -\frac{1}{\pi} Im \sum_i \frac{\langle a|\varphi_i\rangle\langle\varphi_i|a\rangle}{\varepsilon - \varepsilon_i + i\delta}\bigg|_{\delta \to 0^+} = -\frac{1}{\pi} Im\, G_{aa}(\varepsilon) \qquad (12.23)$$

Here, $G_{aa}(\varepsilon)$ is the projection on the adsorbate state of the single particle Green function

$$G(\varepsilon) = \sum_i \frac{|\varphi_i\rangle\langle\varphi_i|}{\varepsilon - \varepsilon_i + i\delta} \qquad (12.24)$$

which is defined by the formal matrix equation

$$(\varepsilon - H + i\delta)G(\varepsilon) = \overline{\overline{I}} \qquad (12.25)$$

To get $n_a(\varepsilon)$, we need the imaginary part of the $|a\rangle$ projection of $G(\varepsilon)$.

Using that $\tilde{\varepsilon} = \varepsilon + i\delta$, Equation (12.25) reads

$$\begin{pmatrix} \tilde{\varepsilon} - \varepsilon_a & -V_{a1} & \cdots & -V_{an} \\ -V_{1a} & \tilde{\varepsilon} - \varepsilon_{11} & & 0 \\ \vdots & \vdots & & \vdots \\ -V_{na} & 0 & \cdots & \tilde{\varepsilon} - \varepsilon_{nn} \end{pmatrix} \begin{pmatrix} G_{aa} & G_{a1} & \cdots & G_{an} \\ G_{1a} & G_{11} & & G_{1n} \\ \vdots & \vdots & & \vdots \\ G_{na} & G_{n1} & \cdots & G_{nn} \end{pmatrix} = \begin{pmatrix} 1 & 0 & \cdots & 0 \\ 0 & 1 & & 0 \\ \vdots & \vdots & & \vdots \\ 0 & 0 & \cdots & 1 \end{pmatrix} \qquad (12.26)$$

Considering only equations involving G_{aa}, we get

$$G_{aa} \cdot (\tilde{\varepsilon} - \varepsilon_a) - \sum_k V_{ak} G_{ka} = 1 \cdots \wedge \quad -V_{ka} G_{aa} + (\tilde{\varepsilon} - \varepsilon_k) \cdot G_{ka} = 0 \qquad (12.27)$$

These equations can be solved to give

$$G_{aa} = \frac{1}{(\varepsilon - \varepsilon_a + i\delta) - \sum_k \dfrac{V_{ak}^2}{(\varepsilon - \varepsilon_k + i\delta)}}\bigg|_{\delta \to 0^+} = \frac{1}{\varepsilon - \varepsilon_a - q(\varepsilon)}\bigg|_{\delta \to 0^+} \qquad (12.28)$$

Here, we have introduced the self-energy $q(\varepsilon) = \Lambda(\varepsilon) - i\Delta(\varepsilon)$.

Solving for the imaginary part of this function, we find that

$$\Delta(\varepsilon) = \pi \sum_k V_{ak}^2 \delta(\varepsilon - \varepsilon_k) \qquad (12.29)$$

Once we have the imaginary part of the complex function for the self-energy, we can get the real part directly using the Kramers–Kronig relations

$$\Lambda(\varepsilon) = \frac{1}{\pi} P \int_{-\infty}^{\infty} \frac{\Delta(\varepsilon')}{(\varepsilon - \varepsilon')} d\varepsilon' \tag{12.30}$$

P denotes the Cauchy principal value, which is a mathematical method that provides a way to assign values to otherwise improper integrals.

Inserting these into the expression for the adsorbate-projected DOS, one obtains

$$n_a(\varepsilon) = -\frac{1}{\pi} Im \, G_{aa}(\varepsilon) = -\frac{1}{\pi} Im \left(\frac{1}{\varepsilon - \varepsilon_a - \Lambda(\varepsilon) + i\Delta(\varepsilon)} \right) \tag{12.31}$$

Multiplying the argument by its complex conjugate in the nominator and denominator and extracting the imaginary part, we get that

$$n_a(\varepsilon) = \frac{1}{\pi} \frac{\Delta(\varepsilon)}{(\varepsilon - \varepsilon_a - \Lambda(\varepsilon))^2 + \Delta(\varepsilon)^2} \tag{12.32}$$

We note that the Newns–Anderson model is a tight-binding model that describes everything in terms of single-electron energy levels. Hence, binding energies obtained from this will not agree with total binding energies calculated using self-consistent DFT. As it turns out, in what is known as the *frozen density* approximation, energy changes induced by small variations in coverage on the surface or variations in the metal are given by

$$\delta E_{ads} = \delta E_{1 \, electron} + \delta E_{ES, A-\Omega} \tag{12.33}$$

This shows that in this approximation the differences in energy between near-similar systems become a sum of single-electron energy differences plus differences in the interatomic electrostatic interactions. From this, one can with some authority claim that there is a sound theoretical background for using the one-electron energy spectra to describe binding energy variations between different systems. In the following, we will justify this expression.

12.4   BOND-ENERGY TRENDS

We will write the total energy of a system of interacting electrons as proposed originally by P. Hohenberg and W. Kohn:

$$E_{tot} = T_{HK}[\rho] + F[\rho]$$
$$= T_{HK}[\rho] + \frac{1}{2} \iint \frac{\rho(\vec{r})\rho(\vec{r}')}{|\vec{r} - \vec{r}'|} d\vec{r} d\vec{r}' + \int v(\vec{r})\rho(\vec{r}) d\vec{r} + E_{nn} + E_{xc}[\rho] \tag{12.34}$$

$F[\rho]$ is seen as the sum of the average electrostatic potential, interactions between the electrons and the nuclei through the external potential $v(\vec{r})$, the interactions between the nuclei, and the exchange–correlation energy $E_{xc}[\rho]$. $T_{HK}[\rho]$ is the kinetic energy of a noninteracting gas of electrons moving in an effective potential. The potential is chosen so that the solutions to the one-electron Schrödinger equation satisfy that the noninteracting system has the same electron density as the real system. Under these conditions, $T_{HK}[\rho]$ can be written as

$$T_{HK}[\rho] = \sum_{occ} \varepsilon_i - \int v_{eff}(\vec{r})\,\rho(\vec{r})\,d\vec{r} \tag{12.35}$$

Even though the above clearly shows that the one-electron energies are insufficient to describe total energies of a system, we shall see that with the right assumption, changes in the one-electron energies can yield changes in bond energies correctly.

Let's consider an arbitrary adsorbate a placed outside a metal surface M. In the following, we want to estimate the change in adsorption energy of the adsorbate when the metal is modified slightly to $\tilde{M}$. The modifications we want to consider are perturbations in the electronic structure induced by, for example, another atom or molecule, which is adsorbed on M in the vicinity of a, or if the metal M is exchanged for another metal close to M in the periodic table.

We are interested in the difference in adsorption energy in the two cases:

$$\begin{aligned}
\delta E_{ads} &= \Delta E[\tilde{M}] - \Delta E[M] \\
&= \left(E_{tot}[\tilde{M}+a] - E_{tot}[\tilde{M}] - E_{tot}[a] \right) - \left(E_{tot}[M+a] - E_{tot}[M] - E_{tot}[a] \right) \quad (12.36) \\
&= E_{tot}[\tilde{M}+a] - E_{tot}[\tilde{M}] - \left(E_{tot}[M+a] - E_{tot}[M] \right)
\end{aligned}$$

We divide space up into two regions: a near-adsorbate region A and a metal region Ω (see Fig. 12.7). In the near-adsorbate region, the change in electron density induced

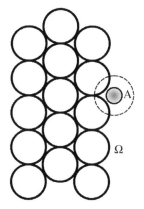

FIGURE 12.7 Schematics showing the two distinct regions (the region near the adsorbate A and the region near the metal Ω) where there is significant perturbation of the system.

by the change of M to $\tilde{M}$ is expected to be very small; hence, the one-electron potential will also only be affected slightly by the change. The same arguments hold true for the metal region, Ω; hence, the effect of the adsorbate on the metal electronic states will be weak.

We can exploit this in connection with the generalized variational principle of DFT, which says that a change in the electron density *and* in the one-electron potential will only give rise to changes in the total energy to the second order. In region A, the dominant electronic effects are set up by the adsorbate a; hence, we choose to apply the same density and potential in this region irrespective of the metal.

Likewise, we let the density and potential in region Ω be independent of the presence of the adsorbate. As a consequence of the variational principle, freezing the density and potential in this way will only result in second-order errors in δE_{ads}.

We begin by considering the contribution of the $F[\rho]$ term to δE_{ads}. In Equation (12.34), $E_{xc}[\rho]$ is mostly described by a local function of position in space. This is not generally true of the electrostatic energy contributions.

One could imagine that the adsorbate has a dipole moment; this will give rise to an electrostatic potential in metal region Ω. For the present, we shall neglect such nonlocal electrostatic interactions between regions A and Ω. In that case, $F[\rho]$ can be divided into contributions from the two regions $F = F_A + F_\Omega$, and therefore, we can write

$$
\begin{aligned}
\delta F_{local} & - F_A[\tilde{M} + a] - F_A[\tilde{M}] - \left(F_A[M + a] - F_A[M] \right) \\
& + F_\Omega[\tilde{M} + a] - F_\Omega[\tilde{M}] - \left(F_\Omega[M + a] - F_\Omega[M] \right) \\
& = \left(F_A[\tilde{M} + a] - F_A[M + a] \right) - \left(F_A[\tilde{M}] - F_A[M] \right) \\
& + \left(F_\Omega[\tilde{M} + a] - F_\Omega[\tilde{M}] \right) - \left(F_\Omega[M + a] - F_\Omega[M] \right)
\end{aligned}
\tag{12.37}
$$

With the assumption that the density in the near-adsorbate region and the metal region can be frozen, all terms in parenthesis in the last equation will be zero. What is left in the adsorption energy difference is the contribution from the nonlocal electrostatic energy from $F[\rho]$:

$$
\delta E_{ES, A-\Omega} = \iint \frac{\rho(\vec{r})\rho(\vec{r}')}{|\vec{r} - \vec{r}'|} \, d\vec{r} d\vec{r}'
\tag{12.38}
$$

In a similar way, the frozen potential and density assumption can be used to show that the net contribution from the $v(\vec{r})\rho(\vec{r})$ integrals in the kinetic energy difference is zero and that only the difference in the one-electron energies calculated with the frozen potentials will contribute to the kinetic energy contribution $T_{HK}[\rho]$:

$$
\begin{aligned}
\delta T_{HK} = \delta E_{1\,electron} & = \sum \varepsilon \left(v_\Omega[\tilde{M}], v_A[a] \right) - \sum \varepsilon \left(v_{\Omega+A}[\tilde{M}] \right) \\
& - \sum \varepsilon \left(v_\Omega[M], v_A[a] \right) - \sum \varepsilon \left(v_{\Omega+A}[M] \right).
\end{aligned}
\tag{12.39}
$$

The difference in adsorption energy is therefore given by the one-electron energy difference plus the difference in electrostatic interaction between the surface and the adsorbate in the two situations, and hence, one obtains the result in Equation (12.44).

The main result from the Newns–Anderson model describes how a single adsorbate state $|a\rangle$ with energy ε_a develops as it approaches a surface with a large number of states $|k\rangle$ $k \in \{1, 2, ..., n\}$. Here, we again use the bra and ket vector notation to describe a specific state of the system.

We shall study a couple of simple cases that will help us understand how different surface band structures affect the adsorbate and hence how strong the adsorbate interacts with the surface. The expression describing the effect on the adsorbate state upon interaction with the surface in its simplest form is given by

$$n_a(\varepsilon) = \frac{1}{\pi} \frac{\Delta(\varepsilon)}{(\varepsilon - \varepsilon_a - \Lambda(\varepsilon))^2 + \Delta(\varepsilon)^2} \qquad (12.40)$$

where $\Delta(\varepsilon)$ can be regarded as a projection of the metal DOS around the adsorbate and the function $\Lambda(\varepsilon)$ is given as a relatively simple transform of $\Delta(\varepsilon)$.

We saw earlier how aluminum had a band structure that looked much like what one would expect for a three-dimensional free electron gas. This is typical for s- and p-bands where the electrons are delocalized. For higher energies, the surface electronic wave functions will have an increasing oscillatory behavior, and hence, the atom-projected DOS will eventually go to zero because the overlapping integrals average out.

In the following, we will approximate the sp electronic band structure with a semielliptic band that initially follows the $\sqrt{\varepsilon}$ behavior and goes to zero for higher energies. When the adsorbate is far from the surface, its electronic distributions are narrow and are best described by delta functions. As it approaches the surface and starts to interact with the delocalized electronic surface states, the spectral distribution will evolve, and during a time length that is characteristic for the specific system, the distribution would have undergone a broadening and a shift (this depends highly on the distribution $\Delta(\varepsilon)$) that is comparable to the dispersion. To see this, we will look at a case where the metal DOS projected around the adsorbate is a constant over the entire energy range: $\Delta(\varepsilon) = \Delta_0$. Later, we will return to the case where $\Delta(\varepsilon)$ is described by a semielliptic distribution. The transformation of $\Delta(\varepsilon) = \Delta_0$ to determine $\Lambda(\varepsilon)$ becomes an integral over an odd function; hence, $\Lambda(\varepsilon) = 0$ and

$$n_a(\varepsilon) = \frac{1}{\pi} \frac{\Delta_0}{(\varepsilon - \varepsilon_a)^2 + \Delta_0^2} \qquad (12.41)$$

This is just an expression for a Lorentzian distribution centered at ε_a and of width Δ_0, and the broadening is simply due to the finite lifetime of the electron in its adsorbate state.

For the semielliptic case, the adsorbate state will undergo the same broadening. But now, due to the finite width of the metal DOS, the maximum of $n_a(\varepsilon)$ will shift downward. The energy value where $n_a(\varepsilon)$ has its maximum is when $\varepsilon - \varepsilon_a - \Lambda(\varepsilon) = 0$,

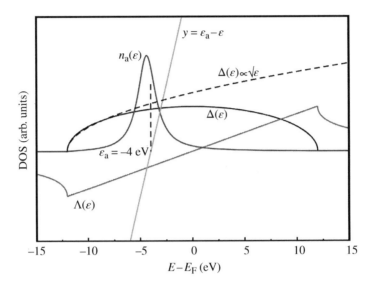

FIGURE 12.8   Plot showing the atom-projected DOS $n_a(\varepsilon)$ for an adsorbate approaching the surface with a single energy level initially at $-4\,eV$ below the Fermi level defined by the surface. The local projection of the surface DOS around the adsorbate level $\Delta(\varepsilon)$ assuming a semielliptic band structure and the corresponding transform $\Lambda(\varepsilon)$ is shown to illustrate why the adsorbate level broadens and shifts down upon interaction with a surface of very delocalized electronic states like the sp electrons. We specifically assume that the surface states behave like a free electron gas for low energies $\Delta(\varepsilon) \propto \sqrt{\varepsilon}$ and that they fall off at higher energies due to the decrease in overlap between the adsorbate state and these surface states.

and the downshift is indicative of stronger bonding. In Figure 12.8, we have shown schematically how this interaction comes about.

Transition metals both have a broad sp-band that leads to the aforementioned broadening and downshift. However, they also have the very localized d electronic states that interact much weaker with the adsorbate than the s and p electronic states. But as we shall see, due to the similarities between the sp-states from one transition metal to the next, the structure of the d-band becomes a very important factor that can help understand differences between the catalytic activities of transition metals.

In Figure 12.9, we have shown, using a semielliptic approach, how a narrow d-band leads to the formation of bonding and antibonding states. The relative density of d-states is much larger than for the sp-states, but due to the close relation between the width of the formed band and the coupling between the adsorbate levels and the surface states, the interaction with the d-band gives less impact on the bond strength than the sp-band does.

As we can see from $\varepsilon - \varepsilon_a - \Lambda(\varepsilon) = 0$, there are now three energy solutions. The lower energy solution is the downshifted bonding state, the middle solution is a weak state that gives no bonding contribution, and finally, we have the high-energy state also known as the upshifted antibonding state. The resemblance with molecular

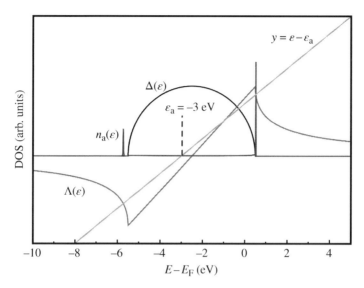

FIGURE 12.9 Plot showing the atom-projected DOS $n_a(\varepsilon)$ for an adsorbate approaching the surface with a single energy level initially at $-3\,eV$ below the Fermi level defined by the surface. The local projection of the surface DOS around the adsorbate level $\Delta(\varepsilon)$ assuming a semielliptic band structure and the corresponding transform $\Lambda(\varepsilon)$ is shown to illustrate why the adsorbate level splits into bonding and antibonding states upon interaction with the surface states. The line is shown to explicitly indicate the three solutions to $d(n_a(\varepsilon))/d\varepsilon = 0$.

orbital theory is striking, and the surface d-states affect the adsorbate energy levels in much the same way as a molecular orbital. Furthermore, relative positions and filling of states are among the factors that determine the strength of the chemical bond formed, which is also the case for molecular orbital theory.

Let us look at the Newns–Anderson model in more detail. We do that by varying the parameters in the model separately. There are three terms that define the bonding between an adsorbate state and a surface in the Newns–Anderson model: (1) the structure of the local projection of the surface DOS around the adsorbate $\Delta(\varepsilon)$, (2) the coupling strength V, and (3) the energy position of the adsorbate level ε_a relative to the Fermi level.

In Figure 12.10a, we have shown explicitly how variations in the width of the local projection of the surface DOS affect the adsorbate-projected DOS. The adsorbate-projected DOS is chosen to be a delta function located at $-5\,eV$ before interaction. The surface DOS is for simplicity assumed to have the form of a semiellipse, and the width is given as the second moment centered at the mean of the semielliptic distribution.

The first plot on the left in Figure 12.10a shows two distinct peaks: one at low energy, which is very much like the delta function of the noninteraction adsorbate state, and one at high energy just at the upper edge of the local projection of the surface DOS. This situation is similar to the coupling of two discrete levels in the weak-coupling limit. The relative strength of the two peaks is due to the large separation between the adsorbate state and the surface states.

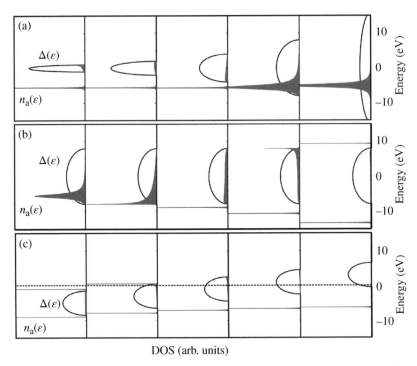

DOS (arb. units)

FIGURE 12.10 Plot showing how the individual relevant parameters in the Newns–Anderson model influence the behavior of the adsorbate-projected DOS. In (a), we vary the width of the local projection of the surface states $\Delta(\varepsilon)$ and keep the coupling strength V and the relative position of the adsorbate level and the surface band fixed. Width increases from left to right. In (b), we see how variations in the coupling strength V affect the bonding behavior for fixed width and relative position. Strength increases left to right. Finally, in (c), we look at variations of the relative position of the distribution and how that affects the bonding for fixed width and coupling strength. In (c), we have included the Fermi level (dashed line) to see explicitly how antibonding states above the Fermi level are depleted as the distance between the adsorbate level and the mean of the band increases. Relative position increases left to right.

As the surface states are broadened (moving to the right in the sequence in Fig. 12.10a), the adsorbate states begin to overlap more and more with the surface states, and the number of states that are pushed up in energy increases. When the bandwidth becomes large enough to actually embed the adsorbate state, the interaction produces a Lorentzian centered at the energy ε_a, thus resembling the interaction between an adsorbate and the delocalized sp electrons, as discussed earlier.

Let us see what happens if we freeze the width of the surface states and the relative position between the adsorbate and the surface states and allow the coupling strength between the states to vary. We choose a situation where the adsorbate DOS is located at the bottom of the band of surface states. Hence, initially, the adsorbate is already interacting with the surface states, but we still see only a single peak. As we increase the coupling strength (moving right in Fig. 12.10b) and $V \sim |\varepsilon_a|$, we see that the adsorbate states get smeared out over the entire surface band. Eventually when $V \gg |\varepsilon_a|$, we are in the strong coupling limit where the interaction approaches the molecular two-level interaction picture, and two distinct peaks are formed above and below the surface DOS.

Finally, we freeze out the coupling strength and the width of the surface states and look at variations due to changes in the separation between the adsorbate state and the surface states. We define the position of the surface states by the statistical mean μ of the distribution. So far, we have used $\mu = 0$, but in the following, we will introduce a Fermi level and set that level as our new zero. Now, the mean relative to the Fermi level and the width of the surface distribution define the filling of the band of surface states; we can explicitly see what happens to the interaction energy as the relative position of the adsorbate and surface states shift. Let's set $\varepsilon_d = \mu$. Now when $|\varepsilon_d - \varepsilon_a|$ increases as the surface states shift to higher energies, as seen when moving right in Figure 12.10c, the number of adsorbate-projected states above the Fermi level decreases. This depletion of the antibonding states will lead to an increase in the bond strength. It reaches a maximum when the band is half filled.

In the next section, we shall combine what we have learned from the Newns–Anderson and tight-binding models and apply that as a basis for understanding the choice of the descriptors that we use to explain trends in surface reactivity.

12.5   BINDING ENERGIES USING THE NEWNS–ANDERSON MODEL

Until now, we have focused mainly on the induced changes on the adsorbate-projected DOS as it interacts with the metal states. We have estimated the bond strength using the molecular orbital theory and found that one can get qualitative agreement with this very simple approach.

Given the value of the coupling strength V, the renormalized energy level ε_a, and the structure $\Delta(\varepsilon)$ and filling f of the band, we can determine the total energy. For the unperturbed metal system, the total energy is given by

$$E_{\text{metal}} = \int_{-\infty}^{E_F} \varepsilon \rho(\varepsilon) d\varepsilon \qquad (12.42)$$

As the adsorbate approaches the surface, it will perturb the system and hence induce a change in the overall density of surface states $\delta\rho(\varepsilon) = (\tilde{\rho}(\varepsilon) - \rho(\varepsilon))$. Here, $\tilde{\rho}(\varepsilon)$ is the density of the metal after the interaction with the adsorbate. The change in the total energy of the system can be obtained from integration of $\varepsilon\delta\rho(\varepsilon)$ over occupied states.

Before we perform the integration, it is, however, important to note that the effect of the perturbation is to introduce some extra states below the Fermi level: δN. These are states that, because of the adsorbate–metal interaction event, have been removed from the Fermi level and introduced into the valence band. The binding energy is now given by

$$E_{bind} = 2 \cdot \int_{-\infty}^{E_F} \varepsilon \delta\rho(\varepsilon) d\varepsilon - 2 \cdot \delta N \cdot E_F + n_a(E_F - \varepsilon_a) \qquad (12.43)$$

Here, we have explicitly added a factor of 2 for spin degeneracy, and we have taken the n_a adsorbate electrons of initial energy ε_a into account by depositing them at the Fermi level and subtracting their energy before bonding.

Choosing $E_F = 0$, the binding energy expression can be written such that

$$E_{bind} = \frac{2}{\pi} \int_{-\infty}^{0} \eta(\varepsilon) d\varepsilon - n_a \varepsilon_a, \qquad (12.44)$$

where $\eta(\varepsilon) = \arctan(\Delta(\varepsilon)/\varepsilon - \varepsilon_a - \Lambda(\varepsilon))$ is an explicit function of the real and imaginary parts of the self-energy expression derived above and given by Equations (12.29) and (12.30).

We can now estimate the bond energy using Equation (12.44), and we can show that the simple d-band model is able to describe variations in bonding. This is seen in Figure 8.5, where the trends in dissociative chemisorption energies for atomic oxygen on a series of $4d$ transition metals are shown. Both experiments and DFT calculations show that the bonding becomes stronger (i.e., ΔE_{ads} becomes more negative) as we move left in the periodic table.

FURTHER READING

Bell AT, Head-Gordon M. Quantum mechanical modeling of catalytic processes. Annu Rev Chem Biomol Eng 2011;2:453.

Hammer B, Nørskov JK. Theoretical surface science and catalysis—calculations and concepts. Adv Catal 2000;45:71.

Hammer B, Morikawa Y, Nørskov JK. CO chemisorption over metal surfaces and overlayers. Phys Rev Lett 1996;76:2141.

Newns DM. Self-consistent model of hydrogen chemisorption. Phys Rev 1969;178:1123.

Nilsson A, Pettersson LGM. Chemical bonding on surfaces probed by X-ray emission spectroscopy and density functional theory. Surf Sci Rep 2004;55 (2–5):49.

INDEX

Activity map, **97–103**, 105–108, 170
Adsorbate–adsorbate interactions, **15**
Adsorption, **7–9**
  entropy, **31–34**
  equilibria, **34–38**
  isotherms, **34–39**
Apparent activation energy, **73**
Arrhenius expression, 6, 10, 47–49, **61–65**, 73, 98, 99, 143

Band diagram, **183**
BEP relations, **91–92**, 140–143, 163
Boltzmann
  constant, **6**
  distribution, **23**, 51–53
  formula, **44**
Born–Oppenheimer approximation, **51**

CatApp, **10–11**, 13, 39–42, 81, 82
Cauchy principle value, **186**
Chemisorption, **7–8**
Configurational entropy, 33, 39, **44–45**, 68, 70

Coupling matrix element, 121–125, **131**, 177–179

d-band
  center, **119–122**, 140, 162
  model, **114–120**, 194
d-projected density of states, **119–120**
Degree of structure sensitivity, **142–144**
Density of states, **116, 119–120**, 181
Diffusion, **13–14**
Dipole moment, **159**

Eigenstates, **176**, 184
Eigenvalues, **176**
Electrochemical cell, **156–157**
Electrostatic potential, **156–157**
Eley–Rideal, **74**
Equilibrium constant, **30–31, 42–43**, 70, 71, 78, 98, 99, 103, 106
Exchange–correlation energy, **187**
External potential, **187**

Fundamental Concepts in Heterogeneous Catalysis, First Edition. Jens K. Nørskov,
Felix Studt, Frank Abild-Pedersen and Thomas Bligaard.
© 2014 John Wiley & Sons, Inc. Published 2014 by John Wiley & Sons, Inc.

Fermi level, **115**
Frozen density approximation, **186**

Gibbs free energy, **26–33**
    diagram, **40–41**

Harmonic transition state theory (HTST),
    49, **61–65**
Heat capacity, **23**
Heisenberg uncertainty principle, **160**
Heyrovsky mechanism, **74**

Interpolation principle, **130**

Kramers–Kronig relations, **186**

Langmuir isotherm, **35–37**, 73
Law of mass action, 30, **42–43**

Mean field model, **69**, 79
Microkinetic modeling, **68–72**
Miller indices, **18**
Minimum energy path, **9**

Near-surface alloys, **127**
Newns–Anderson model, **184–193**
Newtons's 2nd law, **48–50**
Normal hydrogen electrode NHE, **161**

Overpotential, **167–172**
Oxygen
    evolution reaction, **167**
    reduction reaction, 127, **158**

Pauli repulsion, **123–124**
Physisorption, **7–8**
Polarizability, **159–160**

Polarization curve, **165–166**
Potential energy
    diagram, **6–22**
    surface, **9**
Prefactor, 47, **56–58**, 65, 73, 99
Promoter, 138–139, **146–148**

Rate constant, **47–55**, 69–72, 98, 105, 165
Reaction coordinate, **10**, 13, 19, 41
Reversible hydrogen electrode RHE, **161**

Sabatier
    analysis, **103–105**, 110
    map, **104**
    principle, **100**, 103
Schrödinger equation, **176**, 184, 187
Selectivity maps, **110**
Solar fuels, **4**
Steady-state approximation, **76–77**, 97
Stirling's approximation, **45**
Strong metal support interactions SMSI, **148**
Support, 138, 143, **146–148**

Tafel plot, **165**
Tight binding model, **179**, 186, 193
Transition state
    scaling relations, **92**, 99, 100, 106
    theory TST, **49**
Transmission coefficient, **60**, 66
Turnover frequency, **81**

Van der Waals interactions, **7**
Variational principle, **59**
Variational transition state theory, **59**
Volcano relation, **97**

Zero-point energy, **13**, 32, 48, 64